CHEMICAL PRINC
IN THE LABORATORY
With Qualitative Analysis

Alternate Edition

Emil J. Slowinski
Professor of Chemistry
Macalester College
St. Paul, Minnesota

Wayne Wolsey
Associate Professor of Chemistry
Macalester College
St. Paul, Minnesota

William L. Masterton
Professor of Chemistry
University of Connecticut
Storrs, Connecticut

SAUNDERS COLLEGE PUBLISHING
Harcourt Brace Jovanovich College Publishers

Fort Worth Philadelphia San Diego
New York Orlando Austin San Antonio
Toronto Montreal London Sydney Tokyo

Text Typeface: Caledonia
Compositor: York Graphic Services, Inc.
Acquisitions Editor: John Vondeling
Developmental Editor: Lee Walters
Project Editor: Janis Moore
Copy Editors: Janis Moore, Kris Frasch
Managing Editor and Art Director: Richard L. Moore
Design Assistant: Virginia A. Bollard
Text Design: Nancy E. J. Grossman
Cover Design: Richard L. Moore
New Text Artwork: Tom Mallon
Production Manager: Tim Frelick
Assistant Production Manager: Maureen Read

Cover: *Geyser activity in Yellowstone National Park*. Photo © 1982 Richard L. Moore.

CHEMICAL PRINCIPLES IN THE LABORATORY
WITH QUALITATIVE ANALYSIS   ISBN 0-03-062649-8

Requests for permission to make copies of any part of the work should be mailed to: Permissions
Department, Harcourt Brace Jovanovich, Publishers, 8th Floor, Orlando, Florida 32887.
Printed in the United States of America.
Library of Congress catalog card number 82-60632.

34   022   1211

Preface

In spite of its many successful theories, chemistry remains, and probably always will remain, an experimental science. Most of the research in chemistry, both in the universities and in industry, is done in the laboratory rather than in the office or computing room, and it behooves the young student of chemistry to devote a substantial portion of time to the experimental aspects of the subject. It is not easy to become a good experimentalist, and it must be admitted that some chemists are not always effective in the laboratory. Yet even those chemists who do not work full time in the laboratory must be familiar with the available experimental methods, the proper design of experiments, and the interpretation of experimental results. As beginning chemists, students will find that their efforts in the laboratory will be rewarded by a better understanding of the concepts of chemistry as well as an appreciation of what is required in the way of technique and interpretation if one is to be able to find or demonstrate any chemically significant relations.

In writing this manual the authors have attempted to illustrate many of the established principles of chemistry with experiments that are as interesting and challenging as possible. For the most part the experimental procedures and methods for calculation of data are described in great detail, so that students of widely varying backgrounds and abilities will be able to see how to perform the experiments properly and how to interpret them. Several of the experiments in this manual were not published previously and were developed and tested in the general chemistry laboratories at the University of Connecticut and at Macalester College. We have included more experiments than can be conveniently done in the usual laboratory program, so that instructors may select experiments in a flexible way to meet the needs of their particular courses. In many of the experiments unknowns are to be assigned to students to ensure their working independently and to introduce a measure of realism to the application of the chemical principle being investigated.

The choice and order of experiments in this manual were made to be as compatible as possible with our general chemistry text, *Chemical Principles, Alternate Edition*. In that text there is a full section on qualitative analysis, including considerable descriptive chemistry and a complete scheme for the analysis of the common cations and anions. That scheme is the one used in this manual, and it differs from the one in our previous qualitative analysis manual mainly in the procedure for analysis of the Group III cations, which we feel is considerably improved.

In preparing this edition we have made several other minor changes that we hope will be helpful. As before, we include with each experiment an advance study assignment, designed to help the student prepare for the laboratory session. The questions in the advance study assignments include in nearly every case data similar to those the student will obtain in the experiment. Directions for treating the data have been given in great detail, so that if the student properly completes the advance study assignments, he or she should have no trouble in performing the experiments or in processing the data.

We have selected the experiments with some regard to cost, since both chemicals and equipment are expensive. In the teacher's guide we note the cost of the chemicals per student for each experiment. Some of the experiments were modified to lower their cost, but in no case do we believe the value of the experiment was decreased by the changes that were made. We have attempted to keep the experiments safe, have dropped toxic reagents or suspected carcinogens where that was feasible, and have included safety warnings where they seemed necessary. We have switched from hydrogen to natural gas in the reduction of oxides to minimize the risk of an explosion.

With regard to units of volume, we use cubic centimeters (cm^3) in several of the earlier experiments, since that is the volume unit used in much of our text. In most of the manual, however, we use milliliters (ml), since that volume unit is the most common one in the chemical laboratory. Although the milliliter and liter are not, strictly speaking, SI units, they are the volume units most commonly used by chemists, and students should be familiar with them.

We believe that descriptive chemistry is best learned in the laboratory, and that all things considered, qualitative analysis gives students the most interesting introduction to that area of chemistry. The combination of this manual and the text mentioned above offers the instructor the option to include varying amounts of descriptive chemistry in the general course. We hope that these books will be particularly useful in those schools in which general chemistry is a three-quarter course, with the third quarter devoted to qualitative analysis.

The authors would like to acknowledge the assistance of Prof. David A. Katz of the Community College of Philadelphia, who reviewed much of the manuscript and made many helpful suggestions. We are most grateful for the comments and criticisms we have received over the years from users of this manual, faculty and students alike. We assure our readers that we appreciate hearing from them and that we will give careful consideration to any suggestions they would like to make.

E. J. SLOWINSKI
W. C. WOLSEY
W. L. MASTERTON

Safety in the Laboratory

Read This Section Before Performing Any of the Experiments in This Manual

A chemistry laboratory can be, and should be, a safe place in which to work. Yet each year in academic and industrial laboratories accidents occur that in some cases injure seriously, or kill, chemists. Most of these accidents could have been foreseen and prevented, had the chemists involved used the proper judgment and taken proper precautions.

The experiments you will be performing have been selected at least in part because they can be done safely. Instructions in the procedures should be followed carefully and in the order given. Where it seemed likely that a very simple error could produce a dangerous set of conditions, we have noted that error specifically. Sometimes even a change of concentration of one reagent is sufficient to change the conditions of a chemical reaction so as to make it occur in a different way, perhaps at a highly accelerated rate. Do not deviate from the procedure in the manual when performing experiments unless specifically told to do so by your instructor.

One of the simplest, and most important, things you can do to avoid injury in the laboratory is to protect your eyes by routinely wearing safety glasses. Your instructor will tell you what kind of safety glasses or goggles to use. Wear them whenever directed to do so by the procedure or by your instructor.

Chemicals in general are toxic materials. This means that they may act as poisons if they get into your digestive or respiratory system. When you work with chemicals, they will be in containers of one kind or another, but it is easy to spill a liquid or solid on your skin if you are not careful. Some chemicals or reactions produce vapors, which you should avoid inhaling. Some reagents are caustic, which means that they can cause chemical burns on your skin and will eat through clothing. Ordinarily, small amounts of a reagent are readily tolerated by our bodies, but it is only wise to respect any chemicals with which you are not familiar, since they may be more dangerous than they appear to be. As a general rule, avoid getting any chemical on your skin. If that happens, wash it off with plenty of water. Also wash your face and hands when you are through working in the laboratory. You should use similar caution outside the laboratory, since such common materials as gasoline, organic solvents, hair dyes, spray paints, Fiberglas patching resins, and the like are both useful and toxic. Before performing an experiment in this manual, read the procedure through completely. Where a reagent or condition that is potentially hazardous is used, we note the danger in the text or with a CAUTION sign. Be particularly careful when performing experiments of this sort. There aren't many, but there are a few.

The most common accident in the general chemistry laboratory occurs when a student tries to insert glass tubing, a thermometer, or a glass rod into a hole in a rubber stopper. The glass breaks because the student subjects it to excess force in the wrong direction, and sharp glass cuts the student's finger or hand, sometimes severely. Such accidents are completely unnecessary. When you are trying to put a piece of tubing into a rubber stopper, use common sense; we offer the following procedure for your consideration:

1. Make sure the hole in the stopper is only a little bit smaller than the tubing you are working with.
2. Keep both hands close to the stopper when working the tubing into the stopper. Don't hold a foot-long piece of tubing by the end away from the stopper, but rather at a point an inch or two from the stopper.

3. Use a lubricant. A drop or two of glycerine, or even water, will make it much easier to work the tubing into the stopper.

Your laboratory assistant should show you the procedure to be used when you work with tubing and stoppers for the first time. If you do cut yourself go to the instructor, and he or she will decide on the proper treatment.

Although other kinds of accidents are less frequent, they do occasionally occur and should be considered as possibilities. Your volatile organic liquid may ignite if you bring an open flame too close. You may spill a caustic reagent on yourself or your neighbor, or you may somehow get some chemical into your eye or mouth. A common response in such a situation is panic. All too frequently a student will, in the excitement of the incident, do something utterly irrational, such as running screaming from the room when the remedy for the accident is very close at hand. If you see an accident happen to another student, watch for signs of panic and tell the student what to do; if it seems necessary, help him to do it. Call your instructor for assistance. Chemical spills are best handled by washing the area quickly with water from the nearest sink. More severe spills can be treated by using the showers or eye washes that may be available in your laboratory. In case of a fire in a beaker, on the bench, or on your clothing or that of another student, do not panic and run. Smother the fire with an extinguisher, a blanket, or with water, as seems most appropriate at the time. If the fire is in a piece of equipment or on the lab bench and does not appear to require instant action, have your instructor put the fire out. In any event, learn where shower, eye wash, and the fire extinguishers are, so you will not have to look all over if you ever need them in a hurry.

We would like to address a special message to any students for whom English is not a native language. It may be that in some experiments you will be given directions that you do not completely understand. If that happens, *do not* try to do that part of the experiment by simply doing what the student next to you seems to be doing. Ask that student, or your instructor, what the confusing word or phrase means, and when you understand what you should do, go ahead. You will soon learn the language well enough, but until you feel comfortable with it, do not hesitate to ask others to help you with unfamiliar phrases and expressions.

In this laboratory manual we have attempted to describe safe procedures and to employ chemicals that are safe when used properly. Many thousands of students have performed the experiments without having accidents, so you can too. However, we authors cannot be in the laboratory when you carry out the experiments to be sure that you observe the necessary precautions. You and your laboratory supervisor must, therefore, see to it that the experiments are done properly and assume responsibility for any accidents or injuries that may occur.

CONTENTS

QUALITATIVE ANALYSIS

EXPERIMENT

1 • The Densities of Liquids and Solids

One of the fundamental properties of any sample of matter is its density, which is its mass per unit of volume. The density of water is exactly 1.00000 g/cm³ at 4°C and is slightly less than one at room temperature (0.9970 g/cm³ at 25°C). Densities of liquids and solids range from values less than that of water to values considerably greater than that of water. Osmium metal has a density of 22.5 g/cm³ and is probably the densest material known at ordinary pressures.

In any density determination, two quantities must be determined—the mass and the volume of a given quantity of matter. The mass can easily be determined by weighing a sample of the substance on a balance. The quantity we usually think of as "weight" is really the mass of a substance. In the process of "weighing" we find the mass, taken from a standard set of masses, that experiences the same gravitational force as that experienced by the given quantity of matter we are weighing. The mass of a sample of liquid in a container can be found by taking the difference between the mass of the container plus the liquid and the mass of the empty container.

The volume of a liquid can easily be determined by means of a calibrated container. In the laboratory a graduated cylinder is often used for routine measurements of volume. Accurate measurement of liquid volume is made by using a pycnometer, which is simply a container having a precisely definable volume. The volume of a solid can be determined by direct measurement if the solid has a regular geometrical shape. Such is not usually the case, however, with ordinary solid samples. A convenient way to determine the volume of a solid is to measure accurately the volume of liquid displaced when an amount of the solid is immersed in the liquid. The volume of the solid will equal the volume of liquid which it displaces.

In this experiment we will determine the density of a liquid and a solid by the procedure we have outlined. First we weigh an empty flask and its stopper. We then fill the flask completely with water, measuring the mass of the filled stoppered flask. From the difference in these two masses we find the mass of water and then, from the known density of water, we determine the volume of the flask. We empty and dry the flask, fill it with an unknown liquid, and weigh again. From the mass of the liquid and the volume of the flask we find the density of the liquid. To determine the density of an unknown solid metal, we add the metal to the dry empty flask and weigh. This allows us to find the mass of the metal. We then fill the flask with water, leaving the metal in the flask, and weigh again. The increase in mass is that of the added water; from that increase, and the density of water, we calculate the volume of water we added. The volume of the metal must equal the volume of the flask minus the volume of water. From the mass and volume of the metal we calculate its density. The calculations involved are outlined in detail in the Advance Study Assignment.

EXPERIMENTAL PROCEDURE

A. Mass of a Slug. After you are shown how to operate the analytical balance in your laboratory, obtain a numbered metal slug from your instructor. Weigh it on the balance to the nearest 0.001 g. Record the mass and the number of the slug and report it to your instructor. When he has approved your weighing, go to the stockroom and obtain a glass-stoppered flask, which will serve as a pycnometer, and samples of an unknown liquid and an unknown metal.

B. Density of a Liquid. If your flask is not clean and dry, clean it with soap and water, rinse it with a few cubic centimeters of acetone, and dry it by letting it stand for a few minutes in the air or by *gently* blowing compressed air into it for a few moments.

Weigh the dry flask with its stopper on the analytical balance, or the toploading balance if so directed, to the nearest milligram. Fill the flask with distilled water until the liquid level is nearly to the *top* of the ground surface in the neck. Put the stopper in the flask in order to drive out *all* the air and any excess water. Work the stopper gently into the flask, so that it is firmly seated in position. Wipe any water from the outside of the flask with a towel and soak up all excess water from around the top of the stopper.

Again weigh the flask, which should be completely dry on the outside and full of water, to the nearest milligram. Given the density of water at the temperature of the laboratory and the mass of water in the flask, you should be able to determine the volume of the flask very precisely. Empty the flask, dry it, and fill it with your unknown liquid. Stopper and dry the flask as you did when working with the water and then weigh the stoppered flask full of the unknown liquid, making sure its surface is dry. This measurement, used in conjunction with those you made previously, will allow you to find accurately the density of your unknown liquid.

C. Density of a Solid. Pour your sample of liquid from the flask into its container. Rinse the flask with a small amount of acetone and dry it thoroughly. Add small chunks of the metal sample to the flask until the flask is about half full. Weigh the flask, with its stopper and the metal, to the nearest milligram. You should have at least 50 g of metal in the flask.

Leaving the metal in the flask, fill the flask with water and then replace the stopper. Roll the metal around in the flask to make sure that no air remains between the metal pieces. Refill the flask if necessary, and then weigh the dry, stoppered flask full of water plus the metal sample. Properly done, the measurements you have made in this experiment will allow a calculation of the density of your metal sample that will be accurate to about 0.1 per cent.

Pour the water from the flask. Put the metal in its container. Dry the flask and return it with its stopper and your metal sample to the stockroom.

Name _____ Section _____

DATA AND CALCULATIONS: Densities of Liquids and Solids

Metal slug no. _____ Mass of slug _____ g

Unknown liquid no. _____ Unknown solid no. _____

Density of unknown liquid

    Mass of empty flask plus stopper   _____ g

    Mass of stoppered flask plus water   _____ g

    Mass of stoppered flask plus liquid   _____ g

    Mass of water   _____ g

    Volume of flask (density of H_2O at 25°C,
    0.9970 g/cm³; at 20°C, 0.9982 g/cm³)   _____ cm³

    Mass of liquid   _____ g

    Density of liquid   _____ g/cm³

    To how many significant figures can the liquid density
    be properly reported?   _____

Density of unknown metal

    Mass of stoppered flask plus metal   _____ g

    Mass of stoppered flask plus metal plus water   _____ g

    Mass of metal   _____ g

    Mass of water   _____ g

    Volume of water   _____ cm³

    Volume of metal   _____ cm³

    Density of metal   _____ g/cm³

    Would you expect the per cent error in the metal density to be higher or lower than the per cent error in the liquid density as obtained in this experiment? _____

    Why?

ADVANCE STUDY ASSIGNMENT:　Densities of Solids and Liquids

The advance study assignments in this laboratory manual are designed to assist you in making the calculations required in the experiment you will be doing. We do this by furnishing you with sample data and showing in some detail how that data can be used to obtain the desired results. In the advance study assignments we will often include the guiding principles as well as the specific relationships to be employed. If you work through the steps in each calculation by yourself, you should have no difficulty when you are called upon to make the necessary calculations on the basis of the data you obtain in the laboratory.

1.　*Finding the volume of a flask.*　A student obtained a clean dry glass-stoppered flask. She weighed the flask and stopper on an analytical balance and found the total mass to be 32.634 g. She then filled the flask with water and obtained a mass for the full stoppered flask of 59.479 g. From these data, and the fact that at the temperature of the laboratory the density of water was 0.9973 g/cm³, find the volume of the stoppered flask.

　　a.　First we need to obtain the mass of the water in the flask. This is found by recognizing that the mass of a sample is equal to the sum of the masses of its parts. For the filled stoppered flask:

　　　　　Mass of filled stoppered flask = mass of empty stoppered flask + mass of water, so mass of water = mass of filled flask − mass of empty flask

　　　　　mass of water = _____ g − _____ g = _____ g

　　Many mass and volume measurements in chemistry are made by the method used in 1a. This method is called measuring by difference, and is a very useful one.

　　b.　The density of a pure substance is equal to its mass divided by its volume:

$$\text{Density} = \frac{\text{mass}}{\text{volume}} \quad \text{or} \quad \text{volume} = \frac{\text{mass}}{\text{density}}$$

　　The volume of the flask is equal to the volume of the water it contains. Since we know the mass and density of the water, we can find its volume and that of the flask. Make the necessary calculation.

　　　　　　Volume of water = volume of flask = _____ cm³

2.　*Finding the density of an unknown liquid.*　Having obtained the volume of the flask, the student emptied the flask, dried it, and filled it with an unknown whose density she wished to determine. The mass of the stoppered flask when completely filled with liquid was 50.376 g. Find the density of the liquid.

　　a.　First we need to find the mass of the liquid by measuring by difference:

　　　　　Mass of liquid = _____ g − _____ g = _____ g

Continued on following page　　　**5**

b. Since the volume of the liquid equals that of the flask, we know both the mass and volume of the liquid and can easily find its density using the equation in 1b. Make the calculation.

Density of liquid = _____ g/cm^3

3. *Finding the density of a solid.* The student then emptied the flask and dried it once again. To the empty flask she added pieces of a metal until the flask was about half full. She weighed the stoppered flask and its metal contents and found that the mass was 152.047 g. She then filled the flask with water, stoppered it, and obtained a total mass of 165.541 g for the flask, stopper, metal, and water. Find the density of the metal.

a. To find the density of the metal we need to know its mass and volume. We can easily obtain its mass by the method of differences:

Mass of metal = _____ g − _____ g = _____ g

b. To determine the volume of metal, we note that the volume of the flask must equal the volume of the metal plus the volume of water in the filled flask containing both metal and water. If we can find the volume of water, we can obtain the volume of metal by the method of differences. To obtain the volume of the water we first calculate its mass by the method of differences:

Mass of water = mass of (flask + stopper + metal + water)

− mass of (flask + stopper + metal)

Mass of water = _____ g − _____ g = _____ g

The volume of water is found from its density, as in 1b. Make the calculation.

Volume of water = _____ cm^3

c. From the volume of the water we calculate the volume of metal:

Volume of metal = volume of flask − volume of water

Volume of metal = _____ cm^3 − _____ cm^3 = _____ cm^3

From the mass of and volume of metal we find the density, using the equation in 1b. Make the calculation.

Density of metal = _____ g/cm^3

EXPERIMENT

2 • Resolution of Matter into Pure Substances, I. Fractional Crystallization

One of the important problems faced by chemists is that of determining the nature and state of purity of the substances with which they work. In order to perform meaningful experiments, chemists must ordinarily use essentially pure substances, which are often prepared by separation from complex mixtures.

In principle the separation of a mixture into its component substances can be accomplished by carrying the mixture through one or more physical changes, experimental operations in which the nature of the components remains unchanged. Because the physical properties of various pure substances are different, physical changes frequently allow an enrichment of one or more substances in one of the fractions that is obtained during the change. Many physical changes can be used to accomplish the resolution of a mixture, but in this experiment we will restrict our attention to one of the simpler ones in common use, namely, fractional crystallization.

The solubilities of solid substances in different kinds of liquid solvents vary widely. Some substances are essentially insoluble in all known solvents; the materials we classify as macromolecular are typical examples. Most materials are noticeably soluble in one or more solvents. Those substances that we call salts often have very appreciable solubility in water but relatively little solubility in any other liquids. Organic compounds, whose molecules contain carbon and hydrogen atoms as their main constituents, are often soluble in organic liquids such as benzene or carbon tetrachloride.

We also often find that the solubility of a given substance in a liquid is sharply dependent on temperature. Most substances are more soluble in a given solvent at high temperatures than at low temperatures, although there are some materials whose solubility is practically temperature-independent and a few others that become less soluble as temperature increases.

By taking advantage of the differences in solubility of different substances we often find it possible to separate the components of a mixture in essentially pure form.

In this experiment you will be given a sample containing silicon carbide, potassium dichromate, and sodium chloride. Your problem will be to separate this mixture into its component parts, using water as a solvent. Silicon carbide SiC is a macromolecular substance and is insoluble in water. Potassium dichromate $K_2Cr_2O_7$ and sodium chloride NaCl are water soluble ionic substances, with different solubilities at different temperatures, as indicated in Figure 2.1. Sodium chloride exhibits little change in solubility between 0°C and 100°C, whereas the solubility of potassium dichromate increases about 16-fold over that temperature range.

Given a mixture containing roughly equal amounts of SiC, NaCl, and $K_2Cr_2O_7$, the silicon carbide is removed first. This is done by stirring the mixture with water, which brings all the sodium chloride and potassium chromate into solution. The insoluble silicon carbide remains behind and can be filtered off.

At this point you are left with a solution containing NaCl and $K_2Cr_2O_7$ in a rather large amount of water. Some of the water is removed by boiling; the solution is then cooled to 0°C. Since $K_2Cr_2O_7$ is not very soluble at this temperature, most of it crystallizes out of solution. In contrast, NaCl, whose solubility is nearly independent of temperature, stays in solution. By this procedure, called fractional crystallization, the two salts are separated from each other.

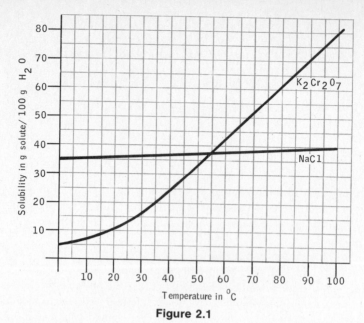

Figure 2.1

EXPERIMENTAL PROCEDURE

WEAR YOUR SAFETY GLASSES WHILE
PERFORMING THIS EXPERIMENT

Obtain from the stockroom a Buchner funnel, a suction flask and a sample (about 25 g) of your unknown solid.

Weigh a 250 ml beaker ($\pm$0.1 g). Add the sample and weigh again. Then add about 100 cm^3 of distilled water, which will be enough to dissolve the soluble solids.

Separation of SiC.   Support the beaker with its solution on a piece of wire gauze over an iron ring, and warm gently to about 40°C. Stir the solution to make sure that all soluble material is dissolved. Remove the insoluble silicon carbide by filtering the solution through a Buchner funnel with gentle suction (see Fig. 2.2). Transfer as much as you can of the solid carbide to the funnel with your rubber policeman. Transfer the orange filtrate to a clean 250 ml beaker. Reassemble the Buchner funnel, apply suction, and wash the SiC on the filter paper with a little distilled water. Continue suction for several minutes to dry the SiC. With the help of your spatula, lift the filter paper and the SiC crystals from the funnel and put the paper on the lab bench so that the crystals may dry in the air.

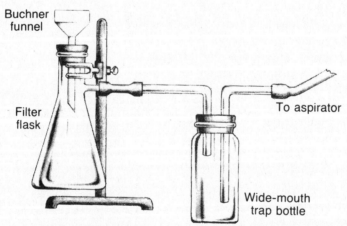

Figure 2.2.   To operate the Buchner funnel, put a piece of circular filter paper in the funnel. Turn on suction and spray filter paper with distilled water from wash bottle. Keep suction on while filtering sample.

Separation of $K_2Cr_2O_7$. Heat the orange filtrate in the beaker to the boiling point and boil gently until white crystals of NaCl are visible in the liquid. The solution will have a tendency to bump, so do not heat it too strongly. *CAUTION: Hot dichromate solution can give you a bad burn.* When NaCl crystals are clearly apparent (the solution will usually appear cloudy at that point), stop heating and add 10 cm³ of distilled water to the solution. (This will be enough to dissolve the NaCl and prevent its crystallizing with the $K_2Cr_2O_7$.) Wash the crystallized solids from the walls of the beaker with your medicine dropper, using the solution in the beaker. Stir the solution with a glass rod to dissolve the solids; if necessary, you may heat the solution, but do not boil it.

Cool the solution to room temperature in a water bath, and then to about 0°C in an ice bath. Bright orange crystals of $K_2Cr_2O_7$ will precipitate. Stir the cold slurry of crystals for several minutes. Assemble the Buchner funnel; chill it by adding 100 cm³ ice-cold distilled water and, after a minute, drawing the water through with suction. Filter the $K_2Cr_2O_7$ slurry through the cold Buchner funnel. Your rubber policeman may be helpful when transferring the last of the crystals. Press the crystals dry with a clean piece of filter paper, and continue to apply suction for a minute or so. Turn off the suction.

Wait a few moments and then, without applying suction, add a few ml of ice-cold distilled water from your wash bottle to the funnel; use just enough water to cover the crystals. Let the cold liquid remain in contact with the crystals for about ten seconds, and then apply suction; the liquid removed will contain most of the NaCl impurity. Continue to apply suction for about a minute to dry the purified $K_2Cr_2O_7$ crystals. Lift the filter paper and the crystals from the funnel and put the paper on the lab bench.

At this point most of the $K_2Cr_2O_7$ in your sample has been separated from the NaCl, which is present in the solution in the suction flask. Although in principle it would be possible to separate the NaCl from the solution we will not attempt to do this, so you may discard the solution in the suction flask.

Analysis of the Purity of the $K_2Cr_2O_7$. The $K_2Cr_2O_7$ crystals you have prepared contain a small amount of NaCl as an impurity. To determine the amount of impurity, weigh out 0.5 ±0.05 g of your recovered $K_2Cr_2O_7$ (it's a little wet, but we won't worry about that) into a weighed 50-ml beaker. Use the top-loading or triple beam balance for the weighing. Dissolve the solid in 10 cm³ of distilled water. Then add 15 drops of 6 M HNO_3, and four drops of 0.1 M $AgNO_3$. Any chloride ion that is present will react with silver ion to form insoluble AgCl and will produce a cloudiness in the solution. Using a stirring rod, mix the reagents with the $K_2Cr_2O_7$ solution and then measure the amount of turbidity in the solution with a spectrophotometer, as directed by your instructor. Determine the amount of NaCl in parts per million from the calibration curve or equation that is provided (1 ppm = 1 g NaCl in 1×10^6 g $K_2Cr_2O_7$).

Recrystallization of $K_2Cr_2O_7$. Clean and dry the 250 ml beaker that you used in the first part of the experiment and add to it your remaining sample of $K_2Cr_2O_7$. Weigh to ±0.1 g on the top loading or triple beam balance. Given the mass of your $K_2Cr_2O_7$ sample, use Figure 2.1 to calculate the minimum amount of water needed to dissolve the sample at 100°C. Add that amount of distilled water to the sample, and heat to boiling. If necessary add a few drops of distilled water to complete the solution process. Cool the solution to room temperature in a water bath and then to 0°C in an ice bath as before. After stirring for several minutes, filter the slurry through an ice-cold Buchner funnel, transferring as much of the solid as possible to the funnel. Press the crystals dry with a piece of filter paper, and continue to apply suction for a minute or two. Lift the filter paper from the funnel and put it on the lab bench. Transfer the solid to a piece of dry filter paper to facilitate drying (your instructor will tell you the mass of the paper).

Determine the amount of NaCl impurity in your recrystallized sample as you did with the previous sample, and record the amount of NaCl present in ppm. The recrystallization should have significantly increased the purity of the $K_2Cr_2O_7$.

Weigh your dry SiC and $K_2Cr_2O_7$ crystals on their pieces of filter paper, using the top loading or triple beam balance. Show your samples of SiC and $K_2Cr_2O_7$ to your laboratory supervisor for evaluation.

Name _____ Section _____

DATA AND CALCULATIONS: Resolution of Pure Substances, I.
Fractional Crystallization

	Original sample	Recrystallized sample
Unknown no. _____		
Mass of 250-ml beaker	_____ g	
Mass of sample and 250-ml beaker	_____ g	
Mass of sample	_____ g	
Mass of 50-ml beaker	_____ g	
Mass of 50-ml beaker plus $K_2Cr_2O_7$	_____ g	_____ g
Mass of $K_2Cr_2O_7$ used in analysis for NaCl	_____ g	_____ g
Parts per million NaCl present in $K_2Cr_2O_7$	_____ ppm	_____ ppm
Mass of SiC plus filter paper	_____ g	
Mass of filter paper (furnished by instructor)	_____ g	_____ g
Mass of SiC in sample	_____ g	
Per cent SiC in sample	_____ %	
Mass of recrystallized $K_2Cr_2O_7$ plus filter paper		_____ g
Mass of recrystallized $K_2Cr_2O_7$		_____ g

Mass of NaCl in the recrystallized $K_2Cr_2O_7$ sample:

$$\text{grams NaCl} = \text{grams } K_2Cr_2O_7 \times \frac{\text{ppm NaCl}}{1 \times 10^6 \text{ g } K_2Cr_2O_7} \qquad \text{_____ g}$$

Name _____ Section _____

ADVANCE STUDY ASSIGNMENT: Resolution of Matter into Pure Substances, I. Fractional Crystallization

1. Using Figure 2.1, determine
    a. the number of grams of $K_2Cr_2O_7$ that will dissolve in 100 grams of H_2O at 100°C.

    _____ g $K_2Cr_2O_7$

    b. the number of grams of water required to dissolve 20 grams of $K_2Cr_2O_7$ at 100°C. Hint: your answer to Part a furnishes you a conversion factor for g $K_2Cr_2O_7$ to g H_2O.

    _____ g H_2O

    c. the number of grams of water required to dissolve 20 grams of NaCl at 100°C.

    _____ g H_2O

    d. the number of grams of water required to dissolve a mixture containing 20 grams of $K_2Cr_2O_7$ and 20 grams of NaCl at 100°C, assuming that the solubility of one substance is not affected by the presence of another.

    _____ g H_2O

2. To the solution in Problem 1.d at 100°C 10 more grams of water are added, and the solution is cooled to 0°C.
    a. How much $K_2Cr_2O_7$ remains in solution?

    _____ g

    b. How much $K_2Cr_2O_7$ crystallizes out?

    _____ g

    c. What per cent of the $K_2Cr_2O_7$ is recovered?

    _____ %

EXPERIMENT

3 • Resolution of Matter into Pure Substances, II. Paper Chromatography

The fact that different substances have different solubilities in a given solvent can be used in several ways to effect a separation of substances from mixtures in which they are present. We have seen in a previous experiment how fractional crystallization allows us to obtain pure substances by relatively simple procedures based on solubility properties. Another widely used resolution technique, which also depends on solubility differences, is chromatography.

In the chromatographic experiment a mixture is deposited on some solid adsorbing substance, which might consist of a strip of filter paper, a thin layer of silica gel on a piece of glass, some finely divided charcoal packed loosely in a glass tube, or even some microscopic glass beads coated very thinly with a suitable adsorbing substance and contained in a piece of copper tubing.

The components of a mixture are adsorbed on the solid to varying degrees, depending on the nature of the component, the nature of the adsorbent, and the temperature. A solvent is then caused to flow through the adsorbent solid under applied or gravitational pressure or by the capillary effect. As the solvent passes the deposited sample, the various components tend, to varying extents, to be dissolved and swept along the solid. The rate at which a component will move along the solid depends on its relative tendency to be dissolved in the solvent and adsorbed on the solid. The net effect is that, as the solvent passes slowly through the solid, the components separate from each other and move along as rather diffuse zones. With the proper choice of solvent and adsorbent, it is possible to resolve many complex mixtures by this procedure. If necessary, we can usually recover a given component by identifying the position of the zone containing the component, removing that part of the solid from the system, and eluting the desired component with a suitable good solvent.

The name given to a particular kind of chromatography depends upon the manner in which the experiment is conducted. Thus, we have column, thin-layer, paper, and vapor chromatography, all in very common use. Chromatography in its many possible variations offers the chemist one of the best methods, if not the best method, for resolving a mixture into pure substances, regardless of whether that mixture consists of a gas, a volatile liquid, or a group of nonvolatile, relatively unstable, complex organic compounds.

In this experiment we will use paper chromatography to separate a mixture of metallic ions in solution. A sample containing a few micrograms of ions is applied as a spot near one edge of a piece of filter paper. That edge is immersed in a solvent, with the paper held vertically. As the solvent rises up the paper by capillary action, it will carry the metallic ions along with it to a degree that depends upon the relative tendency of each ion to dissolve in the solvent and adsorb on the paper. Because the ions differ in their properties, they move at different rates and become separated on the paper. The position of each ion during the experiment can be recognized if the ion is colored, as some of them are. At the end of the experiment their positions are established more clearly by treating the paper with a staining reagent which reacts with each ion to produce a colored product. By observing the position and color of the spot produced by each ion, and the positions of the spots produced by an unknown containing some of those ions, you can readily determine the ions present in the unknown.

It is possible to describe the position of spots such as those you will be observing in terms of a quantity called the R_f value. In the experiment the solvent rises a certain distance, say L centimeters. At the same time a given component will usually rise a smaller distance, say D centimeters. The ratio of D/L is called the R_f value for that component:

$$R_f = \frac{D}{L} = \frac{\text{distance component moves}}{\text{distance solvent moves}} \qquad (1)$$

The R_f value is a characteristic property of a given component in a chromatography experiment conducted under particular conditions. It does not depend upon concentration or upon the other components present. Hence it can be reported in the literature and used by other researchers doing similar analyses. In the experiment you will be doing, you will be asked to calculate the R_f values for each of the cations studied.

EXPERIMENTAL PROCEDURE

Obtain an unknown and a piece of filter paper about 19 cm long by 11 cm wide from the stockroom. Fold the paper in half parallel to the 11-cm edge and then again in thirds so the creases are as shown in Figure 3.1 when you unfold the paper.

Along the 19-cm edge draw a pencil line about 2 cm from that edge. Along the line write the formulas of the cations to be studied, as in Figure 3.1. Put two or three drops of 0.1 M solutions of the following compounds in small micro test tubes, one solution to a tube:

$$Cu(NO_3)_2 \qquad Fe(NO_3)_3$$

$$Ni(NO_3)_2 \qquad Co(NO_3)_2$$

In solution these substances exist as ions. The metallic ions are Cu^{2+}, Ni^{2+}, Fe^{3+}, and Co^{2+}, respectively. One drop of each solution contains about 50 micrograms of cation.

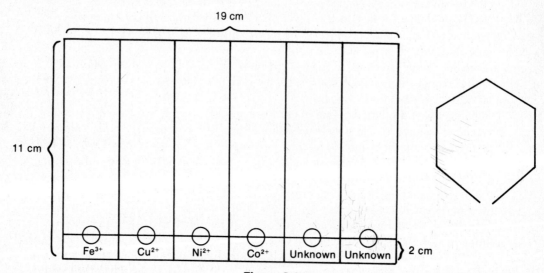

Figure 3.1

For an applicator use a fine capillary tube, which will be furnished to you by your instructor. Test the application procedure by dipping the applicator into one of the solutions and touching it momentarily to a round test piece of filter paper. The liquid from the applicator should form a spot about 5 mm in diameter. Try out this procedure several times.

Clean the applicator by dipping it into a few cm³ of distilled water and blowing air through it to dry it. Dip it into one of the cation solutions and put a 5-mm spot on the line on the filter paper in the region labeled for that cation. Clean the applicator and repeat the procedure with the three other cation-containing solutions and the unknown, putting spots for the unknown in the two areas indicated in Figure 3.1. Dry the spots by moving the paper in the air or by holding it in front of a hair dryer or heat lamp. Then apply the unknown twice more to the same spots. The unknown is less concentrated than the known solutions, so this procedure will increase the amount of each ion. Make sure that the unknown spots are dry before making the next application.

Draw about 15 cm^3 of eluting solution from the supply on the reagent shelf. This solution is made by mixing a solution of HCl, hydrochloric acid, with 2-butanone, an organic solvent. Pour the solution into a 600-ml beaker and cover with a watch glass.

Check to make sure that all the spots on the filter paper are dry. Fold the paper around into a cylindrical hexagon, as in Figure 3.1, using the creases you made originally as the edges of the hexagon. Put the paper into the eluting solution, with the sample spots down near the solution surface. The edges of the paper should *not* touch the wall of the beaker. Re-cover with the watch glass.

The solvent will gradually rise by capillary action on the filter paper, carrying along the different cations at different rates. After the process has gone on for a minute or two you should be able to see colored spots on the paper, showing the positions of some of the cations.

While the experiment is proceeding, you might test the action of the staining reagent on the different cations used in this experiment. Apply a 5-mm spot of each of the cation solutions to a piece of circular filter paper, labeling each spot and cleaning the applicator between solutions. Dry the spots as before. Then, using the spray bottle on the lab bench, spray the paper evenly with the staining reagent, getting the paper moist but not really wet. The staining reagent is a solution containing potassium ferrocyanide and potassium ferricyanide. This reagent forms insoluble colored precipitates with many cations, including all of those used in this experiment. Note the colors obtained with each of the cations.

When the eluting solution has risen to within about 1 cm of the top of the filter paper, remove the paper from the beaker and draw a pencil line along the solvent front. Dry the paper in air or with heat, as before. When the paper is dry, spray with the staining solvent to make the positions of the spots for the known solutions and for the unknown visible. Again dry the paper. From the positions of the various spots, determine which cations are present in your unknown solution. If there is any doubt in your mind about the composition of the unknown, it may be helpful to rinse the filter paper free of staining solution by holding it under the water tap. The spots will not readily wash out of the paper, and their colors are more apparent against a white background. Pour the eluting solution into the waste bottle; don't pour it down the sink.

Measure the distance from the straight pencil line on which you applied the solutions to the solvent front. Then measure the distance from that pencil line to the center of the spot made by each of the cations. Calculate R_f for each cation. Then calculate R_f for each spot formed by the unknown. How do the R_f values compare?

(Optional)   If time permits, you may be interested in seeing if this experiment could be expanded to include more cations. Possible additional cations include

$$Hg^{2+} \quad Cd^{2+} \quad Sb^{3+} \quad Mn^{2+} \quad Ba^{2+} \quad Zn^{2+}$$

In order for a cation to be detected in this experiment, it must meet two criteria. First, it must form a colored product with the staining reagent. Second, it must not move at the same relative speed as one of the other cations studied. Using spots on a piece of circular filter paper determine, after drying, which of the cations listed meet the first criterion. Then, on a piece of folded rectangular paper, carry out the chromatography experiment for those ions which meet the first criterion. Stain the spots, find the R_f values, and decide which, if any, of the ions might be added to the four cations you studied.

EXPERIMENT

4 • Determination of a Chemical Formula

When a substance A reacts with another substance B to form a third substance C, the equation for the chemical reaction can be written as

$$aA + bB \rightarrow cC \tag{1}$$

The substances A and B may be atoms, molecules, or ions in aqueous solution. The numbers a, b, and c are small integers and denote the relative numbers of particles involved in the reaction. Since a mole of any substance contains the same number of particles, be they atoms, molecules, or ions, the numbers a, b, and c also indicate the numbers of moles of A and B that react to form C. There are many reactions that conform to Equation 1, including the following examples:

$$2\ H_2(g) + O_2(g) \rightarrow 2\ H_2O\ (l) \tag{2}$$

$$3\ Ca^{2+}(aq) + 2\ PO_4^{3-}(aq) \rightarrow Ca_3(PO_4)_2(s) \tag{3}$$

Reaction 3 would occur when a solution containing Ca^{2+} ions was mixed with one containing phosphate, PO_4^{3-}, ions. Since the reaction goes essentially to completion there will ordinarily be an excess of one of the reacting ions in the mixture, since the other one will be all used up. If, for example, one slowly added a solution of PO_4^{3-} ions to one containing Ca^{2+} ions, the phosphate ions would react to form $Ca_3(PO_4)_2$ as fast as we added them, so there would, early on, be very little PO_4^{3-} left in solution and an excess of Ca^{2+} ions. As we continued to add PO_4^{3-} ions, more precipitate of $Ca_3(PO_4)_2$ would form, using up Ca^{2+} ions until finally all of the Ca^{2+} initially present would be reacted. After that, further addition of phosphate ion (now in excess) would raise the concentration of that ion, while the concentration of Ca^{2+} ion would remain about zero.

If in carrying out Reaction 3 by the method we describe, we could stop when all of the Ca^{2+} had been converted to $Ca_3(PO_4)_2$, we could verify the formula for $Ca_3(PO_4)_2$ by noting the relative numbers of moles of Ca^{2+} initially in the solution and of PO_4^{3-} that were added. In this case we would find that we needed 2 moles of PO_4^{3-} for every 3 moles of Ca^{2+} in the original solution. This would tell us that the formula of calcium phosphate has to be $Ca_3(PO_4)_2$. In this experiment we will find the chemical formula of an insoluble salt containing metallic cations and chromate, CrO_4^{2-}, anions by using this approach.

In our procedure we will first weigh out a sample of a soluble salt containing a cation that forms an insoluble chromate. An example of such a salt is lead nitrate, $Pb(NO_3)_2$. This salt will serve as our source of metallic cations. Given the mass of the sample, and its formula, we can calculate the number of moles of salt in the sample and, more important, the number of moles of metallic cation it contains. Assuming, say, that we had $Pb(NO_3)_2$ in our sample, and that it weighed 0.4518 grams, we would proceed as follows:

Molar mass $Pb(NO_3)_2$ = (atomic mass Pb + 2 × atomic mass N + 6 × atomic mass O) grams
$$= (207.2 + 2 \times 14.0067 + 6 \times 15.9994)\text{ grams} = 331.2\text{ grams}$$

Number of moles of $Pb(NO_3)_2 = \dfrac{\text{mass of sample}}{\text{molar mass } Pb(NO_3)_2} = \dfrac{0.4518\text{ g}}{331.2\text{ g/mole}} = 1.364 \times 10^{-3}\text{ moles}$

Number of moles of Pb^{2+} = number of moles of $Pb(NO_3)_2 = 1.364 \times 10^{-3}$ moles

Having weighed our sample, we will dissolve it in 20 ml of water, forming just about 20 ml of solution. In the solution the $Pb(NO_3)_2$ will break up, or ionize, completely into Pb^{2+} and NO_3^-

ions. We can calculate the number of moles of Pb^{2+} ions present in one milliliter of solution very easily:

$$\text{Number of moles of } Pb^{2+} \text{ per ml of solution} = \frac{\text{Number of moles of } Pb^{2+}}{\text{Volume of solution in ml}}$$

$$= \frac{1.364 \times 10^{-3} \text{ moles}}{20.0 \text{ ml}} = 6.82 \times 10^{-5} \frac{\text{moles/ml}}{}$$

We add exactly 1 ml of the solution we have prepared to each of 6 small test tubes, numbered 1 to 6.

To test tube No. 1 we then add 1 ml 0.020 M K_2CrO_4. This solution contains 0.020 moles K_2CrO_4 per liter, and since this salt, like all salts, is ionized in solution, it also contains 0.020 moles CrO_4^{2-} per liter, or 2.0×10^{-5} moles CrO_4^{2-} per milliliter. As soon as the Pb^{2+} and CrO_4^{2-} ions meet, they react to form a yellow precipitate of lead chromate. To test tube No. 2 we add 2 ml 0.020 M K_2CrO_4, to test tube No. 3, 3 ml, and so on through test tube No. 6, to which we add 6 ml of the chromate solution.

In some of the test tubes Pb^{2+} ion is in excess, since not enough CrO_4^{2-} ion was added to precipitate all of the cation. In some of the other tubes CrO_4^{2-} ion will be in excess, since more than enough was added to precipitate all of the Pb^{2+} ion. We can determine which ion is in excess in each tube by centrifuging to bring the solid precipitate to the bottom of the tube. The strong yellow color of chromate ion will be apparent in the solutions in those tubes where CrO_4^{2-} is in excess. Where Pb^{2+} is in excess, the solution will be essentially colorless.

If we did this experiment with the sample of $Pb(NO_3)_2$ we used in our example we would find that Mixtures No. 1, 2, and 3 were colorless after centrifuging and that Mixtures No. 4 through 6 were yellow. This allows us to say that in Mixture No. 3 Pb^{2+} was in excess, while in Mixture No. 4 CrO_4^{2-} was in excess. Using the mixtures in those two tubes we can calculate, within limits, the formula for lead chromate. We proceed as follows:

In Mixture No. 3: No. moles $Pb^{2+} = 6.82 \times 10^{-5}$ moles
No. moles $CrO_4^{2-} = 3$ ml $\times 2.0 \times 10^{-5}$ moles/ml $= 6.0 \times 10^{-5}$ moles

Mole ratio, $CrO_4^{2-}:Pb^{2+} = 6.0 \times 10^{-5}$ moles/6.82×10^{-5} moles $= 0.88:1.00$

In Mixture No. 4: No. moles $Pb^{2+} = 6.82 \times 10^{-5}$ moles
No. moles $CrO_4^{2-} = 4$ ml $\times 2.0 \times 10^{-5}$ moles/ml $= 8.0 \times 10^{-5}$ moles

Mole ratio, $CrO_4^{2-}:Pb^{2+} = 8.0 \times 10^{-5}$ moles/6.82×10^{-5} moles $= 1.2:1.00$

If, in Mixtures No. 3 and 4, we assume that *all* the Pb^{2+} and CrO_4^{2-} are present as solid lead chromate, the formulas we would obtain for the compound would be $Pb(CrO_4)_{0.88}$ in Mixture No. 3 and $Pb(CrO_4)_{1.2}$ in Mixture No. 4. The true formula must lie somewhere between these values, since in Mixture No. 3 Pb^{2+} ion is in excess and in Mixture No. 4 CrO_4^{2-} is in excess. Since the mole ratio for $Pb^{2+}:CrO_4^{2-}$ is expected to be a ratio of small integers, we would guess that the ratio might well be 1:1, and that the associated formula for lead chromate would be $PbCrO_4$.

In the optional part of this experiment we will "fine-tune" the amount of CrO_4^{2-} solution we add, getting closer to the situation in which we add just the right amount to precipitate all of the metallic cation. This will allow us to limit the formula to a narrower range than we get with the first set of mixtures.

EXPERIMENTAL PROCEDURE   SEE WARNING AT
                              END OF EXPERIMENT

Part I

From the stockroom obtain a sample of a soluble salt. You will be given its chemical formula at that time.

On an analytical balance, weigh a clean, dry 50-ml beaker to 0.0001 g. Add your sample to the beaker and weigh once again to the same precision.

Using your graduated cylinder, add 20.0 ml of distilled water to the sample in the beaker. Stir with your stirring rod until the sample is dissolved. Then stir for 2 full minutes to ensure that the solution is thoroughly mixed. Pour some 0.020 M K_2CrO_4 from the stock solution into another clean, dry 50-ml beaker until the beaker is about $\frac{2}{3}$ full. Use this solution as your source of chromate ion.

Set up a hot-water bath, consisting of a 250-ml beaker $\frac{2}{3}$ full of water supported on a wire gauze on an iron ring fastened to a ring stand. Begin heating the water with a Bunsen burner.

Number six small test tubes from 1 to 6 and put them in your test tube rack. The test tubes should be clean but not necessarily dry. Using a graduated 5-ml pipet, add exactly 1 ml of your salt solution to each of the test tubes. Your instructor will demonstrate how to use the pipet properly. Be careful when measuring out these small volumes. Then, using a 10-ml graduated pipet, add 1 ml 0.020 M K_2CrO_4 to test tube No. 1, 2 ml to test tube No. 2, 3 ml to test tube No. 3, and 4 ml to test tube No. 4. Refill the pipet and put 5 ml and 6 ml of the chromate solution in test tubes No. 5 and 6, respectively. To the first five test tubes add water from your wash bottle until the volume in each tube is about that in test tube No. 6. The volumes of reagents in each test are summarized in Table 4.1.

TABLE 4.1 COMPOSITION OF REACTION MIXTURES FOR CHROMATE PRECIPITATIONS

Test tube no.	1	2	3	4	5	6
ml salt solution	1	1	1	1	1	1
ml 0.020 M K_2CrO_4	1	2	3	4	5	6
ml water (approx.)	5	4	3	2	1	0
Mixture no.	1	2	3	4	5	6

Shake each tube for at least 30 seconds, using a cork stopper and wiping the cork on a piece of paper towel between tubes. Then put all of the tubes in the hot-water bath, which should be simmering by now. Let the tubes stay in the water bath for about five minutes to promote formation of large crystals of the chromate precipitate. Keep the water bath at the simmering point by judicious adjustment of your Bunsen burner. Remove the tubes from the bath and centrifuge them in batches of two, four, or six tubes, depending on the centrifuge capacity. Then arrange the tubes in order of increasing number from left to right in your test tube rack.

If all goes well, you will find that the solutions above the first few mixtures, with the lower numbers, will be essentially colorless, indicating that the cation is in excess, while the rest of the mixtures will produce yellow solutions, owing to the presence of excess chromate ion. There should be two adjacent tubes in the set of six, one containing a colorless solution and a slight excess of cation and the other yellow with a slight excess of chromate ion. Record the numbers of those two tubes.

Since the two tubes whose numbers you recorded contain mixtures that differ by 1 ml in the amount of chromate solution used to prepare them, we know, within 1 ml, the volume of chromate solution needed to precipitate all of the cation in 1 ml of salt solution. We will call the test tube with the colorless solution Tube A and the one with the yellow solution Tube B.

Part II (Optional)

In this part of the experiment we will attempt to determine to within about 0.2 ml the volume of chromate solution needed to precipitate the cation in 1 ml of the salt solution you prepared. We will do this by adding small but increasing volumes of the chromate solution to six mixtures, all made up identical to the one in Tube A, and detecting the point at which change in solution color occurs, in much the same way as in Part I.

Pour the six mixtures you prepared in Part I into the waste crock on your lab bench. Clean out the test tubes with your test tube brush, rinse the tubes well with distilled water, and put them back in the test tube rack in order of increasing number.

Carefully pipet 1 ml of your salt solution into each of the six tubes, using the 5-ml graduated pipet. Then pipet the same number of ml of the 0.020 K_2CrO_4 solution used to make the mixture in Tube A into each of those six tubes. Essentially you should make six replicas of the mixture that was in Tube A.

Fill a medicine dropper with distilled water from a small beaker, and add the water, drop by drop, to your 10-ml graduated cylinder. Count the number of drops needed to deliver the first ml, the second, and the third, until you are sure you know how many drops it takes to make 1 ml, within a drop or two.

To test tubes No. 1 to 6 we want to add chromate solution in slowly increasing volumes, starting with roughly 0.2 ml in tube No. 1 and ending with about 1 ml in tube No. 6. So, take the number of drops per ml, as delivered by your dropper, and divide by six. If you don't get an integer, take the next larger whole number of drops. (If you have 15 drops per milliliter, you would take 3 drops; if you have 18 drops per ml, you would also take 3 drops.)

Rinse your medicine dropper with the 0.020 M K_2CrO_4 solution in your beaker. Then add the number of drops you just calculated to test tube No. 1, twice that number to test tube No. 2, three times that number to test tube No. 3, and so on, finally adding six times that number to test tube No. 6. In this way, we will get six different mixtures, some with excess cation and some with excess chromate ion, as in Part I. Add distilled water from your wash bottle to each tube to bring the volume up to within about 2 cm of the top of the tube.

Now proceed as you did in Part I, shaking each tube after stoppering it, heating the tubes in the water bath, and centrifuging to bring down the chromate precipitate. Then arrange the tubes in the test-tube rack in order of increasing number.

Once again, there should be a series of tubes with lower numbers above which the solution is colorless, followed by a series with the larger numbers for which the solution is yellow. Since now the difference in composition is not as great as in Part I, the changes in color will not be so great, but you should still be able to select two adjacent tubes, one with essentially a colorless solution and the next with a yellow one. Record the numbers of those two tubes, which we will now call Tubes C and D, respectively.

CAUTION: The solids used in this experiment are toxic. Avoid contact with the crystals or with their solutions. Wash your hands when you finish the experiment.

EXPERIMENT

5 • Law of Multiple Proportions

Many elements form more than one compound with oxygen or other nonmetals. For example, iron forms three oxides, with the formulas FeO, Fe_2O_3, and Fe_3O_4. These oxides, like any set of binary compounds containing the same two elements, can be used to illustrate the Law of Multiple Proportions. If we start with a mole of Fe, the numbers of moles of O combining with that amount of iron are 1, $\frac{3}{2}$, and $\frac{4}{3}$, respectively; all these numbers are either integers or simple fractions, as required by the Law. By analyzing a set of compounds like the oxides of iron we can find their formulas and check the validity of the Law of Multiple Proportions.

In this experiment we will be studying some binary compounds containing copper and bromine. You will be furnished a sample of one of these copper bromides (compound A). On heating a weighed sample of compound A you will find that it breaks down to another copper bromide (compound B), liberating bromine, Br_2, in the process. The reaction might be described as

$$CuBr_x(s) \xrightarrow{\text{heat}} CuBr_y + \frac{(x-y)}{2} Br_2(g) \tag{1}$$
$$\quad \text{A} \qquad\qquad \text{B}$$

We will then convert compound B to an oxide of copper (compound C) by treating it with nitric acid and then heating it. The reactions are fairly complex but the net effect can be represented by the equation:

$$CuBr_y(s) + \frac{z}{2} O_2(g) \rightarrow CuO_z(s) + \frac{y}{2} Br_2(g) \tag{2}$$

The copper oxide is then reduced to copper by heating in the presence of natural gas. By weighing the sample at the end of each step in this series of reactions, we can find the masses of compounds A, B, and C, along with the mass of the copper they contain. From these data, and the atomic masses of copper, bromine, and oxygen, we can find the formulas of the two bromides and the oxide and determine whether the Law of Multiple Proportions applies to copper-bromine compounds.

EXPERIMENTAL PROCEDURE

WEAR YOUR SAFETY GLASSES WHILE
PERFORMING THIS EXPERIMENT

Place about one gram of the copper bromide (A) in a large, accurately weighed test tube. Weigh the test tube and its contents to the nearest 0.001 g. Clamp the test tube so that it is slightly inclined, and incorporate it into an apparatus such as that shown in Figure 5.1. The bent glass tube should extend about an inch past the end of the rubber stopper, and should go to within about an inch of the surface of the solution in the 500-ml Florence flask. *Do not* let the end of the tube dip into the liquid; if it does, liquid will back up into the test tube during the experiment and cause trouble.

Heat the bromide sample, first gently and then rather strongly, with the burner flame. The sample will decompose and bromine, Br_2, will be evolved and absorbed in the solution. *Bromine is a very reactive and very poisonous substance. Do not inhale the vapor or touch the dark liquid.* Continue heating until bromine is no longer produced. (Do not heat the test tube to redness during this step, since at high temperatures the initial decomposition product may be further decomposed to copper.) Let the test tube cool, and remove it from the apparatus. If there is any bromine

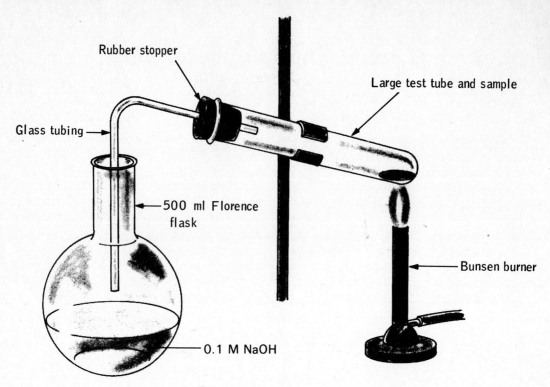

Rubber stopper

Large test tube and sample

Glass tubing

500 ml Florence flask

Bunsen burner

0.1 M NaOH

Figure 5.1

condensed on the walls of the tube, remove it by warming the test tube gently *in the hood* with a Bunsen flame. Allow the tube to cool and then weigh it on the balance.

Add 2 cm³ of 15 M HNO_3 (*CAUTION: CAUSTIC REAGENT*) to the test tube and reassemble the apparatus as before. Heat the test tube gently and then strongly for a few minutes to produce black copper oxide (compound C). Weigh the test tube and its contents when cool.

In the last part of the experiment we will reduce the copper oxide to metallic copper with natural gas. Using the two-holed stopper with the glass tubing inserts, set up the apparatus shown in Figure 5.2. Before incorporating the large test tube into the apparatus, tap the tube gently on the lab bench top to dislodge the caked oxide and break up the solid as much as possible. If necessary, use your stirring rod to break up the solid, being careful to remove any solid adhering to the rod before you take it out of the tube (tap the rod against the tube wall). We will pass the natural gas over the sample and then burn the excess gas in the Bunsen burner. Connect the longer, unbent tubing to the gas jet with a piece of rubber tubing. Use a second piece of rubber tubing to connect the shorter bent tube to your Bunsen burner. Make sure all joints are tight. Have your instructor inspect your apparatus before proceeding further.

Adjust the air control on your Bunsen burner to close off the air supply. Turn on the gas and light the burner. Let the flame burn until it becomes luminous; at this point all the air has been swept from the tube and it is safe to begin the reduction reaction.

Adjust the burner so that the flame is hot and nonluminous, and start heating the sample of copper oxide. As the reduction proceeds the black oxide will gradually take on the color of copper. It will take at least 10 minutes for the reduction to be complete. You should continue heating for about 5 minutes after it appears that the reaction is over, heating any parts of the tube that appear to be in contact with dark oxide.

When you are satisfied that you have carried out the reduction to completion, stop heating and let the tube cool. Do not turn off the burner, however, but just put it to the side. When the tube is cool, turn off the burner and weigh the test tube and its copper contents.

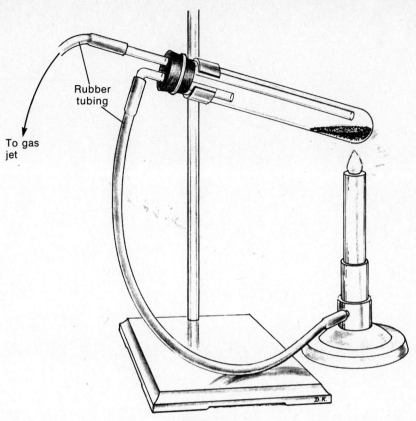

Figure 5.2

EXPERIMENT

6 • Water of Hydration

Most solid chemical compounds will contain some water if they have been exposed to the atmosphere for any length of time. In most cases the water is present in very small amounts, and is merely adsorbed on the surface of the crystals. Other solid compounds contain larger amounts of water that is chemically bound in the crystal. These compounds are usually ionic salts. The water that is present in these salts is called water of hydration and is usually bound to the cations in the salt.

The water molecules in a hydrate are removed relatively easily. In most cases simply heating a hydrate to a temperature somewhat above the boiling point of water will drive off the water of hydration. Hydrated copper(II) chloride is typical in this regard; it is converted to anhydrous $CuCl_2$ if heated to about 110°C:

$$CuCl_2 \cdot 2\ H_2O \rightarrow CuCl_2(s) + 2\ H_2O(g) \text{ at } t \geqslant 110°C$$

In the dehydration reaction the crystal structure of the solid will change and the color of the salt may also change. On heating $CuCl_2 \cdot 2\ H_2O$ the green hydrated crystals are converted to a brownish-yellow powder. You may be familiar with hydrated $CoCl_2$, which is sometimes used in inexpensive hygrometers. $CoCl_2 \cdot 6\ H_2O$ is red, $CoCl_2 \cdot 2\ H_2O$ is violet, and $CoCl_2$ is blue.

Some hydrates lose water to the atmosphere upon standing. This process is called efflorescence. The amount of water lost depends upon the amount of water in the air, as measured by its relative humidity. In moist, warm air, $CoCl_2$ is fully hydrated and is red; in dry cold air $CoCl_2$ loses most of its water of hydration and is blue; at intermediate humidities, $CoCl_2$ exists as a dihydrate and is violet.

Some anhydrous ionic compounds will tend to absorb water from the air or other sources so strongly that they can be used to dry liquids or gases. These substances are called desiccants, and are said to be hygroscopic. A few ionic compounds can take up so much water from the air that they dissolve in the water they absorb; sodium hydroxide, $NaOH$, will do this. This process is called deliquescence.

Some compounds evolve water on being heated but are not true hydrates. The water is produced by decomposition of the compound rather than by loss of water of hydration. Organic compounds, particularly carbohydrates, behave this way. Decompositions of this sort are not reversible; adding water to the product will not regenerate the original compound. True hydrates typically undergo reversible dehydration. Adding water to anhydrous $CuCl_2$ will cause formation of $CuCl_2 \cdot 2\ H_2O$ or, if enough water is added, you will get a solution containing hydrated Cu^{2+} ions. All ionic hydrates are soluble in water, and are usually prepared by crystallization from water solution. The amount of bound water may depend upon the way the hydrate is prepared, but in general the number of moles of water per mole of ionic compound is either an integer or a multiple of $\frac{1}{2}$.

In this experiment you will study some of the properties of hydrates. You will identify the hydrates in a group of compounds, observe the reversibility of the hydration reaction, and test some substances for efflorescence or deliquescence. Finally you will be asked to determine the amount of water lost by a sample of unknown hydrate on heating. From this amount, if given the formula or the molar mass of the anhydrous sample, you will be able to calculate the formula of the hydrate itself.

EXPERIMENTAL PROCEDURE

A. Identification of Hydrates. Place about 0.5 g of each of the compounds listed below in small, dry test tubes, one compound to a tube. Observe carefully the behavior of each compound when you heat it gently with a burner flame. If droplets of water condense on the cool upper walls of the test tube, this is evidence that the compound may be a hydrate. Note the nature and color of the residue. Let the tube cool and try to dissolve the residue in a few cm³ of water, warming very gently if necessary. A true hydrate will tend to dissolve in water, producing a solution with a color very similar to that of the original hydrate. If the compound is a carbohydrate, it will give off water on heating and will tend to char. The solution of the residue in water will often be caramel colored.

Nickel chloride	Sucrose
Potassium chloride	Potassium dichromate
Sodium tetraborate (borax)	Barium chloride

B. Reversibility of Hydration. Gently heat a few crystals, ~0.3 g, of hydrated cobalt(II) chloride, $CoCl_2 \cdot 6\ H_2O$, in an evaporating dish until the color change appears to be complete. Dissolve the residue in the evaporating dish in a few cm³ of water from your wash bottle. Heat the resulting solution to boiling (**CAUTION!**), and carefully boil it to dryness. Note any color changes. Put the evaporating dish on the lab bench and let it cool.

C. Deliquescence and Efflorescence. Place a few crystals of each of the compounds listed below on separate watch glasses and put them next to the dish of $CoCl_2$ prepared in Part B. Depending upon their composition and the relative humidity (amount of moisture in the air), the samples may gradually either lose water of hydration to, or pick up water from, the air. They may also remain unaffected. Any changes in crystal structure, color, or appearance of wetness should be noted. Observe the samples occasionally during the rest of the laboratory period. Since the changes tend to occur slowly, your instructor may have you compare your samples with some that were set out in the laboratory a day or two earlier.

$Na_2CO_3 \cdot 10\ H_2O$ (washing soda)	$KAl(SO_4)_2 \cdot 12\ H_2O$ (alum)
$CaCl_2$	$CuSO_4 \cdot 5\ H_2O$

D. Per Cent Water in a Hydrate. Clean a porcelain crucible and its cover with 6 M HNO_3. Any stains that are not removed by this treatment will not interfere with this experiment. Rinse the crucible and cover with distilled water. Put the crucible with its cover slightly ajar on a clay triangle and heat with a burner flame, gently at first and then to redness for about 2 minutes. Allow the crucible and cover to cool, and then weigh them to 0.001 g on an analytical balance. Handle the crucible with clean crucible tongs.

Obtain a sample of unknown hydrate from the stockroom and place about a gram of sample in the crucible. Weigh the crucible, cover, and sample on the balance. Put the crucible on the clay triangle, with the cover in an off-center position to allow the escape of water vapor. Heat again, gently at first and then strongly, keeping the bottom of the crucible at red heat for about 10 minutes. Center the cover on the crucible and let it cool to room temperature. Weigh the cooled crucible along with its cover and contents.

Examine the solid residue. Add water until the crucible is two thirds full and stir. Warm gently if the residue does not dissolve readily. Does the residue appear to be soluble in water?

Name _____ Section _____

DATA AND OBSERVATIONS: Water of Hydration

A. Identification of Hydrates

	H_2O appears	Color of residue	Water soluble	Hydrate
Nickel chloride	_____	_____	_____	_____
Potassium chloride	_____	_____	_____	_____
Sodium tetraborate	_____	_____	_____	_____
Sucrose	_____	_____	_____	_____
Potassium dichromate	_____	_____	_____	_____
Barium chloride	_____	_____	_____	_____

B. Reversibility of Hydration

Summarize your observations on $CoCl_2 \cdot 6\ H_2O$:

Is the dehydration and hydration of $CoCl_2$ reversible?

C. Deliquescence and Efflorescence

	Observation	Conclusion
$Na_2CO_3 \cdot 10\ H_2O$	_____	_____
$KAl(SO_4)_2 \cdot 12\ H_2O$	_____	_____
$CaCl_2$	_____	_____
$CuSO_4 \cdot 5\ H_2O$	_____	_____

Continued on following page

D.   Per Cent Water in a Hydrate

Mass of crucible and cover   _____ g

Mass of crucible, cover, and solid hydrate   _____ g

Mass of crucible, cover, and residue   _____ g

CALCULATIONS AND RESULTS

Mass of solid hydrate   _____ g

Mass of residue   _____ g

Mass of H_2O lost   _____ g

Percentage H_2O in the unknown hydrate   _____ %

Formula mass of anhydrous salt (if furnished)   _____

Number of moles of water per mole of unknown hydrate   _____

Unknown no.   _____

ADVANCE STUDY ASSIGNMENT: Water of Hydration

1.   A student puts a sample of $Na_2SO_4 \cdot 10\ H_2O$ on a watch glass and observes it occasionally over a period of about an hour. She observes that the crystals gradually change from colorless, transparent, and reasonably large, to a fine white powder. She adds water to the powder and it dissolves. On evaporation of some of the water and subsequent cooling of the solution, large colorless crystals are produced.

   Explain these observations in light of the discussion section of this experiment.

2.   The student is given a sample of a green nickel sulfate hydrate. She weighs the sample in a dry covered crucible and obtains a mass of 22.068 g for the crucible, cover, and sample. The mass of the empty crucible and cover had been found earlier to be 21.214 g. She then heats the crucible to drive off the water of hydration, keeping the crucible at red heat for 10 minutes with the cover slightly ajar. On cooling, she finds the mass of crucible, cover, and contents to be 21.717 g. The sample was converted in the process to yellow solid anhydrous $NiSO_4$.

   a. What is the mass of the hydrate sample?

   _____ g hydrate

   b. What is the mass of the anhydrous $NiSO_4$?

   _____ g $NiSO_4$

   c. What is the mass of water driven off?

   _____ g H_2O

   d. What is the per cent water in the hydrate?

   $$\% \text{ water} = \frac{\text{mass of water in sample}}{\text{mass of hydrate sample}} \times 100\%$$

   _____ %

   e. How many moles of water would there be in 100.0 g of hydrate? (Use the result of Part d to find the number of grams of water in 100.0 g of hydrate, and convert that amount to moles.)

   _____ moles H_2O

Continued on following page   **45**

f. How many moles of $NiSO_4$ are there in 100.0 g of hydrate? (What per cent of the hydrate is $NiSO_4$? Find the number of grams of $NiSO_4$ in 100 g of hydrate. Convert that mass to moles of $NiSO_4$. Formula mass $NiSO_4 = 154.8$.)

_____ moles $NiSO_4$

g. How many moles of water are present per mole of $NiSO_4$?

_____

h. What is the formula of the hydrate?

_____

EXPERIMENT

7 • Heat Effects and Calorimetry

Heat is a form of energy, sometimes called thermal energy, which can pass spontaneously from an object at a high temperature to an object at a lower temperature. If the two objects are in contact they will, given sufficient time, both reach the same temperature.

Heat flow is ordinarily measured in a device called a calorimeter. A calorimeter is simply a container with insulating walls, made so that essentially no heat is exchanged between the contents of the calorimeter and the surroundings. Within the calorimeter chemical reactions may occur or heat may pass from one part of the contents to another, but no heat flows into or out of the calorimeter from or to the surroundings.

A. Specific Heat. When heat flows into a substance the temperature of that substance will increase. The quantity of heat Q required to cause a temperature change Δt of any substance is proportional to the mass m of the substance and the temperature change, as shown in Equation (1). The proportionality constant is called the specific heat of that substance.

$$Q = (\text{specific heat}) \times m \times \Delta t = S.H. \times m \times \Delta t \qquad (1)$$

The specific heat can be considered to be the amount of heat required to raise the temperature of one gram of the substance by one degree Celsius (if you make m and Δt in Equation 1 both equal to one, then Q will equal $S.H.$). Amounts of heat are measured in either joules or calories. To raise the temperature of one gram of water by one degree Celsius 4.18 joules of heat must be added to the water. The specific heat of water is therefore 4.18 joules/g°C. Since 4.18 joules equals one calorie, we can also say that the specific heat of water is 1.00 calories/g°C. Ordinarily heat flow into or out of a substance is determined by the effect which that flow has on a known amount of water. Because water plays such an important role in these measurements the calorie, which was the unit of heat most commonly used until recently, was actually defined to be equal to the specific heat of water.

The specific heat of a metal can readily be measured in a calorimeter. A weighed amount of metal is heated to some known temperature and is then quickly poured into a calorimeter that contains a measured amount of water at a known temperature. Heat flows from the metal to the water, and the two equilibrate at some temperature between the initial temperatures of the metal and the water.

Assuming that no heat is lost from the calorimeter to the surroundings, and that a negligible amount of heat is absorbed by the calorimeter walls, the amount of heat that flows from the metal as it cools is equal to the amount of heat absorbed by the water.

In thermodynamic terms, the heat flow for the metal is equal in magnitude but opposite in direction, and hence in sign, to that for the water. For the heat flow Q,

$$Q_{\text{metal}} = -Q_{\text{H}_2\text{O}} \qquad (2)$$

If we now express heat flow in terms of Equation 1 for both the water and the metal M, we get

$$S.H._M m_M \Delta t_M = -S.H._{\text{H}_2\text{O}} m_{\text{H}_2\text{O}} \Delta t_{\text{H}_2\text{O}} \qquad (3)$$

In this experiment we measure the masses of water and metal and their initial and final temperatures. (Note that $\Delta t_M < 0$ and $\Delta t_{\text{H}_2\text{O}} > 0$, since $\Delta t = t_{\text{final}} - t_{\text{initial}}$.) Given the specific heat of water we can find the positive specific heat of the metal by Equation 3. We will use this procedure to obtain the specific heat of an unknown metal.

The specific heat of a metal is related in a simple way to its atomic mass. Dulong and Petit discovered many years ago that about 25 joules were required to raise the temperature of one mole of many metals by one degree Celsius. This relation, shown in Equation 4, is known as the Law of Dulong and Petit:

$$AM \cong \frac{25}{S.H.(J/g°C)} \tag{4}$$

where AM is the atomic mass of the metal. Once the specific heat of the metal is known, the approximate atomic mass can be calculated by Equation 4. The Law of Dulong and Petit was one of the few rules available to early chemists in their studies of atomic masses.

B. Heat of Solution. When a chemical reaction occurs in water solution, the situation is similar to that which is present when a hot metal sample is put into water. With such a reaction there is an exchange of heat between the reaction mixture and the solvent, water. As in the specific heat experiment, the heat flow for the reaction mixture is equal in magnitude but opposite in sign to that for the water. The heat flow associated with the reaction mixture is also equal to the enthalpy change, ΔH, for the reaction, so we obtain the equation

$$Q_{reaction} = \Delta H_{reaction} = -Q_{H_2O} \tag{5}$$

By measuring the mass of the water used as solvent, and by observing the temperature change that the water undergoes, we can find Q_{H_2O} by Equation 1 and ΔH by Equation 5. If the temperature of the water goes up, heat has been *given off* by the reaction mixture, so the reaction is *exo*thermic; Q_{H_2O} is positive and ΔH is *negative*. If the temperature of the water goes down, the reaction mixture has *absorbed heat from* the water and the reaction is *endo*thermic. In this case Q_{H_2O} is negative and ΔH is *positive*. Both exo- and endothermic reactions are observed.

One of the simplest reactions that can be studied in solution occurs when a solid is dissolved in water. As an example of such a reaction note the solution of NaOH in water:

$$NaOH(s) \rightarrow Na^+(aq) + OH^-(aq); \ \Delta H = \Delta H_{solution} \tag{6}$$

When this reaction occurs, the temperature of the solution becomes much higher than that of the NaOH and water that were used. If we dissolve a known amount of NaOH in a measured amount of water in a calorimeter, and measure the temperature change that occurs, we can use Equation 1 to find Q_{H_2O} for the reaction and use Equation 5 to obtain ΔH. Noting that ΔH is directly proportional to the amount of NaOH used, we can easily calculate $\Delta H_{solution}$ for either a gram or a mole of NaOH. In the second part of this experiment you will measure $\Delta H_{solution}$ for an unknown ionic solid.

EXPERIMENTAL PROCEDURE   WEAR YOUR SAFETY GLASSES WHILE PERFORMING THIS EXPERIMENT

A. Specific Heat. From the stockroom obtain a calorimeter, a sensitive thermometer, a sample of metal in a large stoppered test tube, and a sample of unknown solid. (The thermometer is very expensive, so be careful when handling it.)

The calorimeter consists of two nested expanded polystyrene coffee cups fitted with a styrofoam cover. There are two holes in the cover for a thermometer and a glass stirring rod with a loop bent on one end. Assemble the experimental setup as shown in Figure 7.1.

Fill a 400-cm³ beaker two-thirds full of water and begin heating it to boiling. While the water is heating, weigh your sample of unknown metal in the large stoppered test tube to the nearest 0.1 g on a top loading or triple beam balance. Pour the metal into a dry container and weigh the empty test tube and stopper. Replace the metal in the test tube and put the *loosely* stoppered tube into the hot water in the beaker. The water level in the beaker should be high enough so that the top of the metal is below the water surface. Continue heating the metal in the water for at least 5

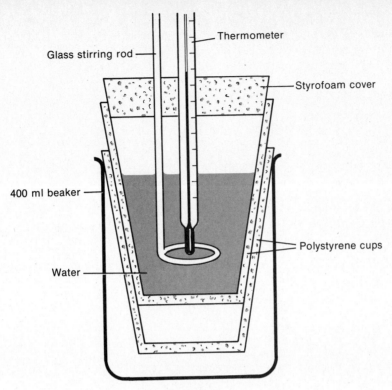

Glass stirring rod

Thermometer

Styrofoam cover

400 ml beaker

Polystyrene cups

Water

Figure 7.1

minutes after the water begins to boil to ensure that the metal attains the temperature of the boiling water. Add water as necessary to maintain the water level.

While the water is boiling, weigh the calorimeter to 0.1 g. Place about 40 cm³ of water in the calorimeter and weigh again. Insert the stirrer and thermometer into the cover and put it on the calorimeter. The thermometer bulb should be completely under the water.

Measure the temperature of the water in the calorimeter to 0.1°C. Take the test tube out of the beaker of boiling water, remove the stopper, and pour the metal into the water in the calorimeter. Be careful that no water adhering to the outside of the test tube runs into the calorimeter when you are pouring the metal. Replace the calorimeter cover and agitate the water as best you can with the glass stirrer. Record to 0.1°C the maximum temperature reached by the water. Repeat the experiment, using about 50 cm³ of water in the calorimeter. Be sure to dry your metal before reusing it; this can be done by heating the metal briefly in the test tube in boiling water and then pouring the metal onto a paper towel to drain. You can dry the hot test tube with a little compressed air.

The metal used in this part of the experiment is to be returned to the stockroom in the test tube in which you obtained it.

B. Heat of Solution. Place about 50 cm³ of distilled water in the calorimeter and weigh as in the previous procedure. Measure the temperature of the water to 0.1°C. The temperature should be within a degree or two of room temperature. In a small beaker weigh out about 5 g of the solid compound assigned to you. Make the weighing of the beaker and of the beaker plus solid to 0.1 g. Add the compound to the calorimeter. Stirring continuously and occasionally swirling the calorimeter, determine to 0.1°C the maximum or minimum temperature reached as the solid dissolves. Check to make sure that all the solid dissolved. A temperature change of at least five degrees should be obtained in this experiment. If necessary, repeat the experiment, increasing the amount of solid used.

　　　　　　　　Name _____ Section _____

DATA AND CALCULATIONS: Calorimetry

A. Specific Heat

	Trial 1	Trial 2
Mass of stoppered test tube plus metal	_____ g $\longrightarrow$	_____ g
Mass of test tube and stopper	_____ g $\longrightarrow$	_____ g
Mass of calorimeter	_____ g $\longrightarrow$	_____ g
Mass of calorimeter and water	_____ g	_____ g
Mass of water	_____ g	_____ g
Mass of metal	_____ g $\longrightarrow$	_____ g
Initial temperature of water in calorimeter	_____ °C	_____ °C
Initial temperature of metal (assume 100°C unless directed to do otherwise)	_____ °C $\longrightarrow$	_____ °C
Equilibrium temperature of metal and water in calorimeter	_____ °C	_____ °C
Δt_{water} ($t_{final} - t_{initial}$)	_____ °C	_____ °C
Δt_{metal}	_____ °C	_____ °C
Specific heat of the metal (Eq. 3)	_____ J/g°C	_____ J/g°C
Approximate atomic mass of metal	_____	_____
Unknown no.	_____	

B. Heat of Solution

Mass of calorimeter plus water	_____ g
Mass of beaker	_____ g
Mass of beaker plus solid	_____ g

Continued on following page　　　**51**

Mass of water, m_{H_2O} _____ g

Mass of solid, m_S _____ g

Original temperature   _____ °C

Final temperature   _____ °C

Q_{H_2O} for the reaction (Eq. 1) (S.H. = 4.18 J/g°C)

_____ joules

ΔH for the reaction (Eq. 5)

_____ joules

The quantity you have just calculated is approximately° equal to the heat of solution of your sample. Calculate the heat of solution per gram of solid sample.

$$\Delta H_{solution} = \text{_____ joules/g}$$

The solution reaction is endothermic exothermic. (Encircle correct answer.) Give your reasoning.

Solid unknown no. _____

(Optional) Formula of compound used (if furnished by instructor) _____

Mass of one mole of compound   _____ g

Heat of solution per mole of compound   _____ joules

° The value of ΔH will be approximate for several reasons. One of them is that we do not include the amount of heat absorbed by the solute. This effect is smaller than the likely experimental error, and thus we will ignore it.

ADVANCE STUDY ASSIGNMENT: Heat Effects and Calorimetry

1.   A metal sample weighing 63.2 g and at a temperature of 100°C was placed in 41.0 g of water in a calorimeter at 24.5°C. At equilibrium the temperature of the water and metal was 35.0°C.

   a. What was Δt for the water? ($\Delta t = t_{final} - t_{initial}$)

_____ °C

   b. What was Δt for the metal?

_____ °C

   c. Taking the specific heat of water to be 4.18 J/g°C, calculate the specific heat of the metal, using Eq. 3.

_____ joules/g°C

   d. What is the approximate atomic mass of the metal? (Use Eq. 4.)

_____

2.   When 2.0 g of NaOH were dissolved in 49.0 g water in a calorimeter at 24.0°C, the temperature of the solution went up to 34.5°C.
   a. Is this solution reaction exothermic? _____ Why?

   b. Calculate Q_{H_2O}, using Eq. 1.

_____ joules

   c. Find ΔH for the reaction as it occurred in the calorimeter (Eq. 5).

$\Delta H = $ _____ joules

   d. Find ΔH for the solution of 1.00 g NaOH in water.

$\Delta H = $ _____ joules/g

   e. Find ΔH for the solution of one mole NaOH in water.

$\Delta H = $ _____ joules/mole

Continued on following page   53

f. Given that NaOH exists as Na^+ and OH^- ions in solution, write the equation for the reaction that occurs when NaOH is dissolved in water.

g. Using enthalpies of formation as given in thermodynamic tables, calculate ΔH for the reaction in Part f and compare your answer with the result you obtained in Part e.

EXPERIMENT

8 • Analysis of an Aluminum-Zinc Alloy*

Some of the more active metals will react readily with solutions of strong acids, producing hydrogen gas and a solution of a salt of the metal. Small amounts of hydrogen are commonly prepared by the action of hydrochloric acid on metallic zinc:

$$Zn(s) + 2\,H^+(aq) \rightarrow H_2(g) + Zn^{2+}(aq) \tag{1}$$

From this equation it is clear that one mole of zinc produces one mole of hydrogen gas in this reaction. If the hydrogen were collected under known conditions, it would be possible to calculate the mass of zinc in a pure sample by measuring the amount of hydrogen it produced on reaction with acid.

Since aluminum reacts spontaneously with strong acids in a manner similar to that shown by zinc,

$$2\,Al(s) + 6\,H^+(aq) \rightarrow 2\,Al^{3+}(aq) + 3\,H_2(g) \tag{2}$$

we could find the amount of aluminum in a pure sample by measuring the amount of hydrogen produced by its reaction with an acid solution. In this case two moles of aluminum would produce three moles of hydrogen.

Since the amount of hydrogen produced by a gram of zinc is not the same as the amount produced by a gram of aluminum,

$$1\ mole\ Zn \rightarrow 1\ mole\ H_2,\ 65.4\ g\ Zn \rightarrow 1\ mole\ H_2,\ 1.00\ g\ Zn \rightarrow 0.0153\ mole\ H_2 \tag{3}$$

$$2\ moles\ Al \rightarrow 3\ moles\ H_2,\ 54.0\ g\ Al \rightarrow 3\ moles\ H_2,\ 1.00\ g\ Al \rightarrow 0.0556\ mole\ H_2 \tag{4}$$

it is possible to react an alloy of zinc and aluminum of known mass with acid, determine the amount of hydrogen gas evolved, and calculate the percentages of zinc and aluminum in the alloy, using relations (3) and (4). The object of this experiment is to make such an analysis.

In this experiment you will react a weighed sample of an aluminum-zinc alloy with an excess of acid and collect the hydrogen gas evolved over water (Fig. 8.1). If you measure the volume, temperature, and total pressure of the gas and use the Ideal Gas Law, taking proper account of the pressure of water vapor in the system, you can calculate the number of moles of hydrogen produced by the sample:

$$P_{H_2}V = n_{H_2}RT, \quad n_{H_2} = \frac{P_{H_2}V}{RT} \tag{5}$$

The volume V and the temperature T of the hydrogen are easily obtained from the data. The pressure exerted by the dry hydrogen P_{H_2} requires more attention. The total pressure P of gas in the flask is, by Dalton's Law, equal to the partial pressure of the hydrogen P_{H_2} plus the partial pressure of the water vapor P_{H_2O}:

$$P = P_{H_2} + P_{H_2O} \tag{6}$$

*W. L. Masterton, J. Chem. Educ. 38, 558 (1961).

The water vapor in the flask is present with liquid water, so the gas is saturated with water vapor; the pressure P_{H_2O} under these conditions is equal to the vapor pressure VP_{H_2O} of water at the temperature of the experiment. This value is constant at a given temperature, and will be found in Appendix I at the end of this manual. The total gas pressure P in the flask is very nearly equal to the barometric pressure P_{bar}.[*]

Substituting these values into (6) and solving for P_{H_2}, we obtain

$$P_{H_2} = P_{bar} - VP_{H_2O} \tag{7}$$

Using (5), you can now calculate n_{H_2}, the number of moles of hydrogen produced by your weighed sample. You can then calculate the percentages of Al and Zn in the sample by properly applying (3) and (4) to your results. For a sample containing g_{Al} grams Al and g_{Zn} grams Zn, it follows that

$$n_{H_2} = (g_{Al} \times 0.0556) + (g_{Zn} \times 0.0153) \tag{8}$$

For a one gram sample, g_{Al} and g_{Zn} represent the mass fractions of Al and Zn, that is, % Al/100 and % Zn/100. Therefore

$$N_{H_2} = \left(\frac{\% \text{ Al}}{100} \times 0.0556\right) + \left(\frac{\% \text{ Zn}}{100} \times 0.0153\right) \tag{9}$$

where N_{H_2} = number of moles of H_2 produced *per gram* of sample.

Since it is also true that

$$\% \text{ Zn} = 100 - \% \text{ Al} \tag{10}$$

(9) can be written in the form

$$N_{H_2} = \left(\frac{\% \text{ Al}}{100} \times 0.0556\right) + \left(\frac{100 - \% \text{ Al}}{100} \times 0.0153\right) \tag{11}$$

We can solve equation (11) directly for % Al if we know the number of moles of H_2 evolved per gram of sample. To save time in the laboratory and to avoid arithmetic errors, it is highly desirable to prepare in advance a graph giving N_{H_2} as a function of % Al. Then when N_{H_2} has been determined in the experiment, % Al in the sample can be read directly from the graph. Directions for preparing such a graph are given in Problem 1 in the Advance Study Assignment.

EXPERIMENTAL PROCEDURE

WEAR YOUR SAFETY GLASSES WHILE
PERFORMING THIS EXPERIMENT

Obtain a suction flask, large test tube, stopper assemblies and a sample of Al-Zn alloy from the stockroom. Assemble the apparatus as shown in Figure 8.1.

Take a gelatin capsule from the supply on the lab bench and weigh it on the analytical balance to ±0.0001 g. Pour your alloy sample out on a piece of paper and add about half of it to the capsule. If necessary, break up the turnings into smaller pieces by simply tearing them. Cover the capsule and weigh it again. The mass of sample should be between 0.1000 and 0.2000 grams. Use care in both weighings, since the sample is small and a small weighing error will produce a large experimental error. Put the remaining alloy back in its container.

Fill the suction flask and beaker about ⅔ full of water. Moisten the stopper on the suction flask and insert it firmly into the flask. Open the pinch clamp and apply suction to the tubing attached

---

[*] In principle a small correction should be made for the difference in heights of the water levels inside and outside the suction flask. In practice the error made by neglecting this effect is much smaller than other experimental errors.

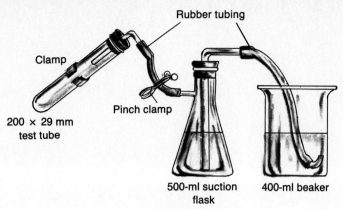

Figure 8.1

to the side arm of the suction flask. Pull water into the flask from the beaker until the water level in the flask is about 4 or 5 cm below the side arm. To apply suction, use a suction bulb or a short piece of rubber tubing attached temporarily to the tube that goes through the test tube stopper. Close the pinch clamp to prevent siphoning. The tubing from the beaker to the flask should be full of water, with no air bubbles.

Carefully remove the tubing from the beaker and put the end on the lab bench. As you do this, no water should leak out of the end of the tubing. Pour the water remaining in the beaker into another beaker, letting the 400-ml beaker drain for a second or two. Without drying it, weigh the empty beaker on a top-loading balance to ±0.1 g. Put the tubing back in this beaker.

Pour 10 ml of 6 M HCl, hydrochloric acid, as measured in your graduated cylinder, into the large test tube. Drop the gelatin capsule into the HCl solution; if it sticks to the tube, poke it down into the acid with your stirring rod. Insert the stopper firmly into the test tube and open the pinch clamp. If a little water goes into the beaker at that point, pour that water out, letting the beaker drain for a second or two.

Within 3 or 4 minutes the acid will eat through the wall of the capsule and begin to react with the alloy. The hydrogen gas that is formed will go into the suction flask and displace water from the flask into the beaker. The volume of water that is displaced will equal the volume of gas that is produced. As the reaction proceeds you will probably observe a dark foam, which contains particles of unreacted alloy. The foam may carry some of the alloy up the tube. Wiggle the tube gently to make sure that all of the alloy gets into the acid solution. The reaction should be over within five to ten minutes. At that time the liquid solution will again be clear, the foam will be essentially gone, the capsule will be all dissolved, and there should be no unreacted alloy. When the reaction is over, close the pinch clamp and take the tubing out of the beaker. Weigh the beaker and the displaced water to ±0.1 g. Measure the temperature of the water and the barometric pressure.

Pour the acid solution into the sink. Reassemble the apparatus and repeat the experiment with the remaining sample of alloy.

Name _____ **Section** _____

DATA: Analysis of an Aluminum-Zinc Alloy

	Trial 1	Trial 2
Mass of gelatin capsule	_____ g	_____ g
Mass of alloy sample plus capsule	_____ g	_____ g
Mass of empty beaker	_____ g	_____ g
Mass of beaker plus displaced water	_____ g	_____ g
Barometric pressure	_____ mm Hg	
Temperature	_____ °C	

CALCULATIONS

	Trial 1	Trial 2
Mass of alloy sample	_____ g	_____ g
Mass of displaced water	_____ g	_____ g
Volume of displaced water ($d = 1.00$ g/ml)	_____ ml	_____ ml
Volume of H_2, V	_____ liters	_____ liters
Temperature of H_2, T	_____ K	
Vapor pressure of water at T, VP_{H_2O}, from Appendix I	_____ mm Hg	
Pressure of dry H_2, P_{H_2}	_____ mm Hg;	_____ atm
Moles H_2 from sample, n_{H_2}	_____ moles	_____ moles
Moles H_2 per gram of sample, N_{H_2}	_____ moles/g	_____ moles/g
%Al (read from graph)	_____ %	_____ %
Unknown no.	_____	

Name _____ Section _____

ADVANCE STUDY ASSIGNMENT: Analysis of an Aluminum-Zinc Alloy

1.   On the following page, construct a graph of N_{H_2} vs % Al. To do this, refer to Equation 11 and the discussion preceding it. Note that a plot of N_{H_2} vs % Al should be a straight line. (Why?) To fix the position of a straight line it is necessary to locate only two points. The most obvious way to do this is to find N_{H_2} when % Al = 0 and when % Al = 100. If you wish you may calculate some intermediate points (for example, N_{H_2} when % Al = 50, or 20, or 70). All of these points should lie on the same straight line.

2.   A student obtained the following data in this experiment. Fill in the blanks in the data and make the indicated calculations:

Mass of gelatin capsule   0.1168 g   Temperature, t 21°C

Mass of capsule plus alloy
sample   0.2754 g   Temperature, T _____ K

Mass of alloy sample, m _____ g   Barometric pressure   746 mm Hg

Mass of empty beaker   141.2 g   Vapor pressure of
                                                    H_2O at t (Appendix I)   _____ mm Hg

Mass of beaker plus displaced
water   307.7 g
                                                    Pressure of dry H_2,
Mass of displaced water   _____ g P_{H_2} (Eqn 7)   _____ mm Hg

Volume of displaced water   Pressure of dry H_2 _____ atm
(density = 1.00 g/ml)   _____ ml

Volume, V, of H_2 = Volume of displaced water _____ ml; _____ liters

Find the number of moles of H_2 evolved, n_{H_2} (Eqn 5; V in liters, P_{H_2} in atm, T in K, $R = 0.0821$ liter-atm/mole K).

_____ moles H_2

Find N_{H_2}, the number of moles of H_2 per gram of sample (n_{H_2}/m). _____ moles H_2/g

Find the % Al in the sample from the graph prepared for Problem 1. _____ %Al

Find the % Al in the sample by using Equation 11.   _____ %Al

Analysis of an Aluminum-Zinc Alloy
(Advance Study Assignment)

N_{H_2}

0.050

0.040

0.030

0.020

0.010

25% 50% 75% 100%

Per cent Al

9 • Molecular Mass of a Volatile Liquid

One of the important applications of the Ideal Gas Law is found in the experimental determination of the molecular masses of gases and vapors. In order to measure the molecular mass of a gas or vapor we need simply to determine the mass of a given sample of the gas under known conditions of temperature and pressure. If the gas obeys the Ideal Gas Law,

$$PV = nRT \tag{1}$$

If the pressure P is in atmospheres, the volume V in liters, the temperature T in K, and the amount n in moles, then the gas constant R is equal to 0.0821 liter atm/(mole K).

The number of moles n is equal to the mass g of the gas divided by its molar mass in grams (GMM). Substituting into (1), we have

$$PV = \frac{gRT}{GMM} \qquad\qquad GMM = \frac{gRT}{PV} \tag{2}$$

This experiment involves measuring the molar mass of a volatile liquid by using Equation (2). A small amount of the liquid is introduced into a weighed flask. The flask is then placed in boiling water, where the liquid will vaporize completely, driving out the air and filling the flask with vapor at barometric pressure and the temperature of the boiling water. If we cool the flask so that the vapor condenses, we can measure the mass of the vapor and calculate a value for GMM.

EXPERIMENTAL PROCEDURE*   WEAR YOUR SAFETY GLASSES WHILE PERFORMING THIS EXPERIMENT

Obtain a special round-bottomed flask, a stopper and cap, and an unknown liquid from the storeroom. Support the flask on an evaporating dish or in a beaker at all times. If you should break or crack the flask, report it to your instructor immediately so that it can be repaired. With the stopper loosely inserted in the neck of the flask, weigh the empty dry flask on the analytical balance. Use a copper loop, if necessary, to suspend the flask from the hook supporting the balance pan.

Pour about half your unknown liquid, about 5 ml, into the flask. Assemble the apparatus as shown in Figure 9.1. Place the cap on the neck of the flask. Add a few boiling chips to the water in the 600 ml beaker and heat the water to the boiling point. Watch the liquid level in your flask; the level should gradually drop as vapor escapes through the cap. After all the liquid has disappeared and no more vapor comes out of the cap, continue to boil the water gently for 5 to 8 minutes. Measure the temperature of the boiling water. Shut off the burner and wait until the water has stopped boiling (about $\frac{1}{2}$ minute) and then loosen the clamp holding the flask in place. Slide out the flask, remove the cap, and *immediately* insert the stopper used previously.

Remove the flask from the beaker of water, holding it by the neck, which will be cooler. Immerse the flask in a beaker of cool water to a depth of about 5 cm. After holding the flask in the water for about two minutes to allow it to cool, carefully remove the stopper *for not more than a second or two* to allow air to enter, and again insert the stopper. (As the flask cools the vapor inside condenses and the pressure drops, which explains why air rushes in when the stopper is removed.)

---

*See W. L. Masterton and T. R. Williams. J. Chem. Educ. *36*, 528 (1959).

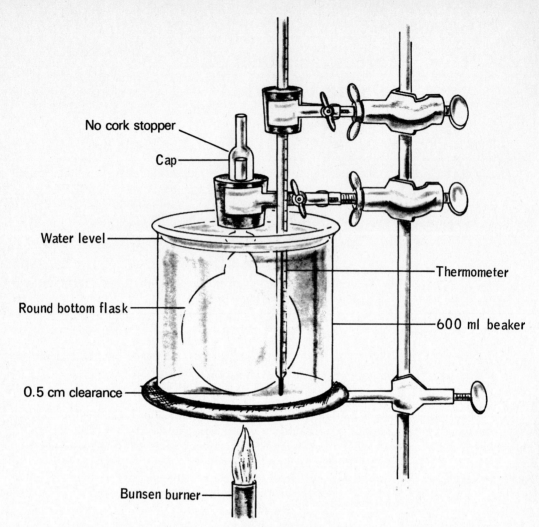

No cork stopper

Cap

Water level

Thermometer

Round bottom flask

600 ml beaker

0.5 cm clearance

Bunsen burner

Figure 9.1

Dry the flask with a towel to remove the surface water. Loosen the stopper momentarily to equalize any pressure differences, and reweigh the flask. Read the atmospheric pressure from the barometer.

Repeat the procedure using another 5 ml of your liquid sample.

You may obtain the volume of the flask from your instructor. Alternatively, he may direct you to measure its volume by weighing the flask stoppered and full of water on a rough balance. *Do not* fill the flask with water unless specifically told to do so.

When you have completed the experiment, return the flask to the storeroom; do not attempt to wash or clean it in any way.

Name _____ Section _____

DATA: Molecular Mass of a Volatile Liquid

	Trial 1	Trial 2
Unknown no.	_____	
Mass of flask and stopper	_____ g	_____ g
Mass of flask, stopper, and condensed vapor	_____ g	_____ g
Mass of flask, stopper, and water (see directions)	_____ g	_____ g
Temperature of boiling water bath	_____ °C	_____ °C
Barometric pressure	_____ mm Hg	_____ mm Hg

CALCULATIONS AND RESULTS

	Trial 1	Trial 2
Pressure of vapor, P	_____ atm	_____ atm
Volume of flask (volume of vapor), V	_____ lit	_____ lit
Temperature of vapor, T	_____ K	_____ K
Mass of vapor, g	_____ g	_____ g
Molar mass of unknown, as found by substitution into Equation 2	_____ g	_____ g

ADVANCE STUDY ASSIGNMENT: Molecular Mass of a Volatile Liquid

1.   A student weighs an empty flask and stopper and finds the mass to be 55.441 g. She then adds about 5 ml of an unknown liquid and heats the flask in a boiling water bath at 100°C. After all the liquid is vaporized, she removes the flask from the bath, stoppers it, and lets it cool. After it is cool, she momentarily removes the stopper, then replaces it and weighs the flask and condensed vapor, obtaining a mass of 56.039 g. The volume of the flask is known to be 215.8 ml. The barometric pressure in the laboratory that day is 752 mm Hg.

   a. What was the pressure of the vapor in the flask in atm?

$$P = \text{_____} \text{atm}$$

   b. What was the temperature of the vapor in K? the volume of the flask in liters?

$$T = \text{_____} \text{K} \qquad V = \text{_____} \text{l}$$

   c. What was the mass of vapor that was present in the flask?

$$g = \text{_____} \text{grams}$$

   d. What is the mass of one mole of vapor? (Eq. 2.)

$$GMM = \text{_____} \text{grams/mole}$$

   e. What is the molecular mass of the unknown?

$$MM = \text{_____}$$

2.   How would each of the following procedural errors affect the results to be expected in this experiment? Give your reasoning in each case.
   a. All of the liquid was not vaporized when the flask was removed from the water bath.

   b. The flask was not dried before the final weighing with the condensed vapor inside.

   c. The flask was left open to the atmosphere while it was being cooled, and the stopper was inserted just before the final weighing.

   d. The flask was removed from the bath before the vapor had reached the temperature of the boiling water.

EXPERIMENT

10 • The Atomic Spectrum of Hydrogen

When atoms are excited, either in an electric discharge or with heat, they tend to give off light. The light is emitted only at certain wavelengths which are characteristic of the atoms in the sample. These wavelengths constitute what is called the atomic spectrum of the excited element and reveal much of the detailed information we have regarding the electronic structure of atoms.

Atomic spectra are interpreted in terms of quantum theory. According to this theory, atoms can exist only in certain states, each of which has an associated fixed amount of energy. When an atom changes its state, it must absorb or emit an amount of energy which is just equal to the difference between the energies of the initial and final states. This energy may be absorbed or emitted in the form of light. The emission spectrum of an atom is obtained when excited atoms fall from higher to lower energy levels. Since there are many such levels, the atomic spectra of most elements are very complex.

Light is absorbed or emitted by atoms in the form of photons, each of which has a specific amount of energy, ϵ. This energy is related to the wavelength of light by the equation

$$\epsilon_{photon} = \frac{hc}{\lambda} \tag{1}$$

where h is Planck's constant, 6.62618×10^{-34} joule seconds, c is the speed of light, 2.997925×10^8 meters per second, and λ is the wavelength, in meters. The energy ϵ_{photon} is in joules and is the energy given off by one atom when it jumps from a higher to a lower energy level. Since total energy is conserved, the change in energy of the atom, $\Delta\epsilon_{atom}$, must equal the energy of the photon emitted:

$$\Delta\epsilon_{atom} = \epsilon_{photon} \tag{2}$$

where $\Delta\epsilon_{atom}$ is equal to the energy in the upper level minus the energy in the lower one. Combining Equations 1 and 2, we obtain the relation between the change in energy of the atom and the wavelength of light associated with that change:

$$\Delta\epsilon_{atom} = \epsilon_{upper} - \epsilon_{lower} = \epsilon_{photon} = \frac{hc}{\lambda} \tag{3}$$

The amount of energy in a photon given off when an atom makes a transition from one level to another is very small, of the order of 1×10^{-19} joules. This is not surprising since, after all, atoms are very small particles. To avoid such small numbers, we will work with one mole of atoms, much as we do in dealing with energies involved in chemical reactions. To do this we need only to multiply Equation 3 by Avogadro's number, N:

Let
$$N\Delta\epsilon = \Delta E = N\epsilon_{upper} - N\epsilon_{lower} = E_{upper} - E_{lower} = \frac{Nhc}{\lambda}$$

Substituting the values for N, h, and c, and expressing the wavelength in nanometers rather than meters (1 meter = 1×10^9 nanometers), we obtain an equation relating energy change in kilojoules per mole of atoms to the wavelength of photons associated with such a change:

$$\Delta E = \frac{6.02205 \times 10^{23} \times 6.62618 \times 10^{-34} \text{ J sec} \times 2.997925 \times 10^8 \text{ m/sec}}{\lambda(\text{in nm})} \times \frac{1 \times 10^9 \text{ nm}}{1 \text{ m}} \times \frac{1 \text{ kJ}}{1000 \text{ J}}$$

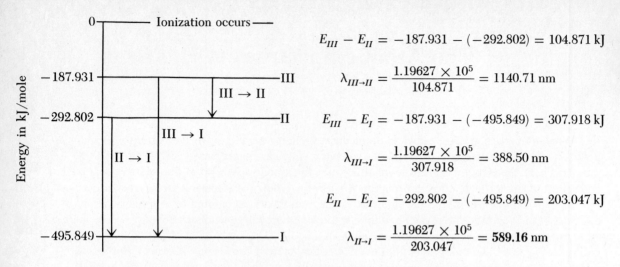

$$E_{III} - E_{II} = -187.931 - (-292.802) = 104.871 \text{ kJ}$$

$$\lambda_{III \to II} = \frac{1.19627 \times 10^5}{104.871} = 1140.71 \text{ nm}$$

$$E_{III} - E_{I} = -187.931 - (-495.849) = 307.918 \text{ kJ}$$

$$\lambda_{III \to I} = \frac{1.19627 \times 10^5}{307.918} = 388.50 \text{ nm}$$

$$E_{II} - E_{I} = -292.802 - (-495.849) = 203.047 \text{ kJ}$$

$$\lambda_{II \to I} = \frac{1.19627 \times 10^5}{203.047} = \mathbf{589.16 \text{ nm}}$$

Figure 10.1. Calculation of wavelengths of spectral lines from energy levels of the sodium atom. I is the ground state, and II and III are excited states.

$$\Delta E = E_{\text{upper}} - E_{\text{lower}} = \frac{1.19627 \times 10^5 \text{ kJ/mole}}{\lambda \text{ (in nm)}} \qquad \text{or} \qquad \lambda \text{ (in nm)} = \frac{1.19627 \times 10^5}{\Delta E \text{ (in kJ/mole)}} \qquad (4)$$

Equation 4 is useful in the interpretation of atomic spectra. Say, for example, we study the atomic spectrum of sodium and find that the wavelength of the strong yellow line is 589.16 nm (see Fig. 10.1). This line is known to result from a transition between two of the three lowest levels in the atom. The energies of these levels are shown in the figure. To make the determination of the levels which give rise to the 589.16 nm line, we note that there are three possible transitions, shown by downward arrows in the figure. We find the wavelengths associated with those transitions by first calculating ΔE ($E_{\text{upper}} - E_{\text{lower}}$) for each transition. Knowing ΔE we calculate λ by Equation 4. Clearly, the II → I transition is the source of the yellow line in the spectrum.

The simplest of all atomic spectra is that of the hydrogen atom. In 1886 Balmer showed that the lines in the spectrum of the hydrogen atom had wavelengths that could be expressed by a rather simple equation. Bohr, in 1913, explained the spectrum on a theoretical basis with his famous model of the hydrogen atom. According to Bohr's theory, the energies allowed to a hydrogen atom are given by the equation

$$\epsilon_n = \frac{-B}{n^2} \qquad (5)$$

where B is a constant predicted by the theory and n is an integer, 1, 2, 3, . . . , called a quantum number. It has been found that all the lines in the atomic spectrum of hydrogen can be associated with energy levels in the atom which are predicted with great accuracy by Bohr's equation. When we write Equation 5 in terms of a mole of H atoms, and substitute the numerical value for B, we obtain

$$E_n = \frac{-1312.04}{n^2} \quad \text{kilojoules per mole,} \quad n = 1, 2, 3, \ldots \qquad (6)$$

Using Equation 6 you can calculate, very accurately indeed, the energy levels for hydrogen. Transitions between these levels give rise to the wavelengths in the atomic spectrum of hydrogen. These wavelengths are also known very accurately. Given both the energy levels and the wavelengths, it is possible to determine the actual levels associated with each wavelength. In this experiment your task will be to make determinations of this type for the observed wavelengths in the hydrogen atomic spectrum that are listed in Table 10.1.

TABLE 10.1   SOME WAVELENGTHS (IN nm) IN THE SPECTRUM OF THE HYDROGEN ATOM AS MEASURED IN A VACUUM

Wavelength	Assignment $n_{hi} \longrightarrow n_{lo}$	Wavelength	Assignment $n_{hi} \longrightarrow n_{lo}$	Wavelength	Assignment $n_{hi} \longrightarrow n_{lo}$
97.25	_____	410.29	_____	1005.2	_____
102.57	_____	434.17	_____	1094.1	_____
121.57	_____	486.27	_____	1282.2	_____
389.02	_____	656.47	_____	1875.6	_____
397.12	_____	954.86	_____	4052.3	_____

EXPERIMENTAL PROCEDURE

There are several ways we might analyze an atomic spectrum, given the energy levels of the atom involved. A simple and effective method is to calculate the wavelengths of some of the lines arising from transitions between some of the lower energy levels, and see if they match those that are observed. We shall use this method in our experiment. All the data are good to at least five significant figures, so by using electronic calculators you should be able to make very accurate determinations.

A.   Calculations of the Energy Levels of the Hydrogen Atom.   Given the expression for E_n in Equation 6, it is possible to calculate the energy for each of the allowed levels of the H atom starting with $n = 1$. Using your calculator, calculate the energy in kJ/mole of each of the ten lowest levels of the H atom. Note that the energies are all negative, so that the *lowest* energy will have the *largest* allowed negative value. Enter these values in the table of energy levels, Table 10.2. On the energy level diagram provided, plot along the y axis each of the six lowest energies, drawing a horizontal line at the allowed level and writing the value of the energy alongside the line near the y axis. Write the quantum number associated with the level to the right of the line.

B.   Calculation of the Wavelengths of the Lines in the Hydrogen Spectrum.   The lines in the hydrogen spectrum all arise from jumps made by the atom from one energy level to another. The wavelengths in nm of these lines can be calculated by Equation 4, where ΔE is the difference in energy in kJ/mole between any two allowed levels. For example, to find the wavelength of the spectral line associated with a transition from the $n = 2$ level to the $n = 1$ level, calculate the difference, ΔE, between the energies of those two levels. Then substitute ΔE into Equation 4 to obtain this wavelength in nanometers.

Using the procedure we have outlined, calculate the wavelengths in nm of all the lines we have indicated in Table 10.3. That is, calculate the wavelengths of all the lines that can arise from transitions between any two of the six lowest levels of the H atom. Enter these values in Table 10.3.

C.   Assignment of Observed Lines in the Hydrogen Spectrum.   Compare the wavelengths you have calculated with those which are listed in Table 10.1. If you have made your calculations properly, your wavelengths should match, within the error of your calculation, several of those which are observed. On the line opposite each wavelength in Table 10.1, write the quantum numbers of the upper and lower states for each line whose origin you can recognize by comparison of your calculated values with the observed values. On the energy level diagram, draw a vertical arrow pointing down (light is emitted, $\Delta E < 0$) between those pairs of levels which you associate with any of the observed wavelengths. By each arrow write the wavelength of the line originating from that transition.

There are a few wavelengths in Table 10.1 which have not yet been calculated. By assignments already made and by an examination of the transitions you have marked on the diagram, deduce the quantum states that are likely to be associated with the as yet unassigned lines. This is perhaps most easily done by first calculating the value of ΔE, which is associated with a given wavelength. Then find two values of E_n whose difference is equal to ΔE. The quantum numbers for the two E_n states whose energy difference is ΔE will be the ones which are to be assigned to the given wavelength. When you have found n_{hi} and n_{lo} for a wavelength, write them in Table 10.1; continue until all the lines in the table have been assigned.

D.   The Balmer Series.   This is the most famous series in the atomic spectrum of hydrogen. Carry out calculations in connection with this series as directed in the Data and Calculations section.

DATA AND CALCULATIONS: The Atomic Spectrum of Hydrogen

A. The Energy Levels of the Hydrogen Atom

Energies are to be calculated from Equation 6 for the ten lowest energy states.

TABLE 10.2

Quantum Number, n	Energy, E_n, in kJ/mole	Quantum Number, n	Energy, E_n, in kJ/mole
___	_____	___	_____
___	_____	___	_____
___	_____	___	_____
___	_____	___	_____
___	_____	___	_____

B. Calculation of Wavelengths in the Spectrum of the H Atom

TABLE 10.3

$$\Delta E = E_{n_{hi}} - E_{n_{lo}}$$

$$\lambda \ (\text{nm}) = \frac{1.19627 \times 10^5}{\Delta E}$$

In the upper half of each box write ΔE, the difference in energy in kJ/mole between $E_{n_{hi}}$ and $E_{n_{lo}}$. In the lower half of the box, write λ in nm associated with that value of ΔE.

Continued on following page

C. Assignment of Wavelengths

1. As directed in the procedure, assign n_{hi} and n_{lo} for each wavelength in Table 10.1 which corresponds to a wavelength calculated in Table 10.3.

2. List below any wavelengths you cannot yet assign and find their origin.

Wavelength λ observed	ΔE transition in kJ/mole	Probable transition $n_{hi} \longrightarrow n_{lo}$	λ calculated in nm (Eq. 4)
_____	_____	_____	_____
_____	_____	_____	_____
_____	_____	_____	_____
_____	_____	_____	_____

D. 1. THE BALMER SERIES. When Balmer found his famous series for hydrogen in 1886, he was limited experimentally to wavelengths in the visible and near ultraviolet regions from 250 nm to 700 nm, so all the lines in his series lie in that region. On the basis of the entries in Table 10.3 and the transitions on your energy level diagram, what common characteristic do the lines in the Balmer Series have?

What would be the longest possible wavelength for a line in the Balmer series?

$$\lambda = \text{_____ nm}$$

What would be the shortest possible wavelength that a line in the Balmer series could have? Hint: What is the largest possible value of ΔE to be associated with a line in the Balmer series?

$$\lambda = \text{_____ nm}$$

Fundamentally, why do all the lines in the hydrogen spectrum between 250 nm and 700 nm belong to the Balmer series? Hint: On the energy-level diagram note the range of values of ΔE for transitions to the $n = 1$ level and for transitions to the $n = 3$ level. Compare those values to ΔE when $\lambda = 250$ nm and when $\lambda = 700$ nm.

2. THE IONIZATION ENERGY OF HYDROGEN. In the normal hydrogen atom the electron is in its lowest energy state, which is called the ground state of the atom. The maximum electronic energy that a hydrogen atom can have is 0 kJ/mole, at which point the electron would essentially be removed from the atom and it would become a H^+ ion. How much energy in kilojoules per mole does it take to ionize an H atom?

$$\text{_____ kJ/mole}$$

Continued on following page

The ionization energy of hydrogen is often expressed in units other than kJ/mole. What would it be in joules per atom?

_____J/atom

(The energy level diagram in Part A is on the following page.)

The Atomic Spectrum of Hydrogen
(Energy Level Diagram)

(DATA AND CALCULATIONS)

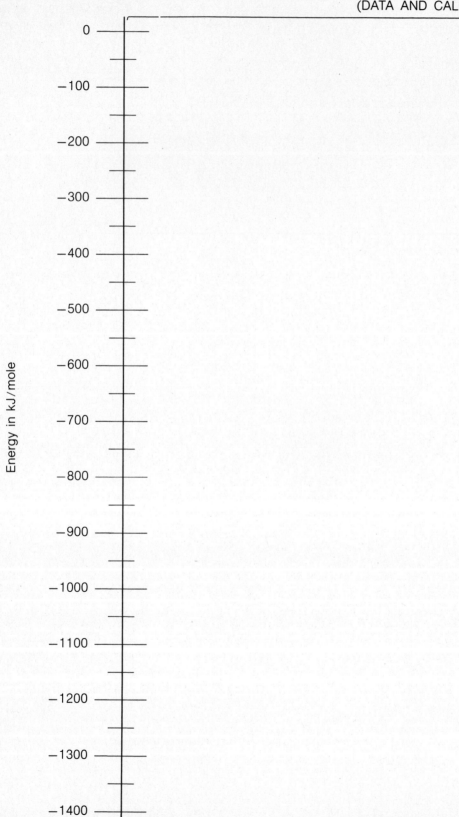

Energy in kJ/mole

ADVANCE STUDY ASSIGNMENT: The Atomic Spectrum of Hydrogen

1. The helium ion, He^+, has energy levels which are similar to those of the hydrogen atom, since both species have only one electron. The energy levels of the He^+ ion are given by the equation

$$E_n = -\frac{5248.16}{n^2} \text{ kJ/mole} \qquad n = 1, 2, 3, \ldots$$

a. Calculate the energies in kJ/mole for the four lowest energy levels of the He^+ ion.

$E_1 = $ _____ kJ/mole

$E_2 = $ _____ kJ/mole

$E_3 = $ _____ kJ/mole

$E_4 = $ _____ kJ/mole

b. One of the most important transitions for the He^+ ion involves a jump from the $n = 2$ to the $n = 1$ level. ΔE for this transition equals $E_2 - E_1$, where these two energies are obtained as in Part a. Find the value of ΔE in kJ/mole. Find the wavelength in nm of the line emitted when this transition occurs; use Equation 4 to make the calculation.

$\Delta E = $ _____ kJ/mole; $\lambda = $ _____ nm

c. Three of the strongest lines in the He^+ ion spectrum are observed at the following wavelengths: (1) 121.57 nm; (2) 164.12 nm; (3) 468.90 nm. Find the quantum numbers of the initial and final states for the transitions which give rise to these three lines. Do this by calculating, using Equation 4, the wavelengths of lines which can originate from transitions involving any two of the four lowest levels. You calculated one such wavelength in Part b. Make similar calculations with the other possible pairs of levels. When a calculated wavelength matches an observed one, write down n_{hi} and n_{lo} for that line. Continue until you have assigned all three of the lines. Make your calculations on the other side of this page.

(1) _____ → _____   (2) _____ → _____   (3) _____ → _____

EXPERIMENT

11 • The Alkaline Earths and the Halogens—Two Families in the Periodic Table

The Periodic Table arranges the elements in order of increasing atomic number in horizontal rows of such length that elements with similar properties recur periodically; that is, they fall directly beneath each other in the Table. The elements in a given vertical column are referred to as a family or group. The physical and chemical properties of the elements in a given family change gradually as one goes from one element in the column to the next. By observing the trends in properties the elements can be arranged in the order in which they appear in the Periodic Table. In this experiment we will study the properties of the elements in two families in the Periodic Table, the alkaline earths (Group 2) and the halogens (Group 7).

The alkaline earths are all moderately reactive metals and include barium, beryllium, calcium, magnesium, radium, and strontium. (Since beryllium compounds are rarely encountered and often very poisonous, and radium compounds are highly radioactive, we will not include these two elements in this experiment.) All the alkaline earths exist in their compounds and in solution as M^{2+} cations (Mg^{2+}, Ca^{2+}, etc.). If solutions of these cations are mixed with solutions containing X^{2-} anions (CO_3^{2-}, SO_4^{2-}, etc.), salts of the alkaline earths will precipitate if the compound MX is insoluble:

$$M^{2+}(aq) + X^{2-}(aq) \rightarrow MX(s) \qquad \text{if MX is insoluble} \qquad (1)$$

$$M^{2+} = Ba^{2+}, Ca^{2+}, Mg^{2+}, \text{ or } Sr^{2+}; X^{2-} = SO_4^{2-}, CO_3^{2-}, C_2O_4^{2-}, \text{ or } CrO_4^{2-}$$

We would expect, and indeed observe, that the solubilities of the salts of the alkaline earth cations with any one of the given anions show a smooth trend consistent with the order of the cations in the Periodic Table. That is, as we go from one end of the alkaline earth family to the other, the solubilities of, say, the sulfate salts either gradually increase or decrease. Similar trends exist for the carbonates, oxalates, and chromates formed by those cations. By determining such trends in this experiment, you will be able to confirm the order of the alkaline earths in the Periodic Table.

The elementary halogens are also relatively reactive. They include astatine, bromine, chlorine, fluorine, and iodine. We will not study astatine and fluorine in this experiment, since the former is radioactive and the latter is too reactive to be safe. Unlike the alkaline earths, the halogen atoms tend to gain electrons, forming X^- anions (Cl^-, Br^-, etc.). Because of this property, the halogens are oxidizing agents, species which tend to oxidize (remove electrons from) other species. An interesting and simple example of the sort of reaction that may occur arises when a solution containing a halogen (Cl_2, Br_2, I_2) is mixed with a solution containing a halide ion (Cl^-, Br^-, I^-). Taking X_2 to be the halogen, and Y^- to be a halide ion, the following reaction may occur:

$$X_2(aq) + 2\ Y^-(aq) \rightarrow 2\ X^-(aq) + Y_2(aq) \qquad (2)$$

The reaction will occur if X_2 is a better oxidizing agent than Y_2, since then X_2 can produce Y_2 by removing electrons from the Y^- ions. If Y_2 is a better oxidizing agent than X_2, Reaction 2 will not proceed but will be spontaneous in the opposite direction.

In this experiment we will mix solutions of halogens and halide ions to determine the relative oxidizing strengths of the halogens. These strengths show a smooth variation as one goes from one halogen to the next in the Periodic Table. We will be able to tell if a reaction occurs by the colors we observe. In water, and particularly in some organic solvents, the halogens have characteristic colors. The halide ions are colorless in water solution and insoluble in organic solvents. Bromine

(Br_2) in 1,1,1-trichloroethane, CCl_3CH_3(TCE) is orange, while Cl_2 and I_2 in that solvent have quite different colors.

Say, for example, we shake a water solution of Br_2 with a little trichloroethane, which is heavier than and insoluble in water. The Br_2 is much more soluble in TCE than in water and goes into the TCE layer, giving it an orange color. To that mixture we add a solution containing a halide ion, say Cl^- ion, and mix well. If Br_2 is a better oxidizing agent than Cl_2, it will take electrons from the chloride ions and will be converted to bromide, Br^-, ions; the reaction would be

$$Br_2(aq) + 2\ Cl^-(aq) \rightarrow 2\ Br^-(aq) + Cl_2(aq) \qquad (3)$$

If the reaction occurs, the color of the TCE layer will of necessity change, since Br_2 will be used up and Cl_2 will form. The color of the TCE layer will go from orange to that of a solution of Cl_2 in TCE. If the reaction does *not* occur, the color of the TCE layer will remain orange. By using this line of reasoning, and by working with the possible mixtures of halogens and halide ions, you should be able to arrange the halogens in order of increasing oxidizing power, which must correspond to their order in the Periodic Table.

One difficulty that you may have in this experiment involves terminology rather than actual chemistry. You must learn to distinguish the halogen *elements* from the halide *ions*, since the two kinds of species are not at all the same, even though their names are similar:

Elementary Halogens	Halide Ions
Bromine, Br_2	Bromide ion, Br^-
Chlorine, Cl_2	Chloride ion, Cl^-
Iodine, I_2	Iodide ion, I^-

The *halogens* are molecular substances and oxidizing agents, and all have odors. They are only slightly soluble in water and are much more soluble in TCE, where they have distinct colors. The *halide ions* exist in solution only in water, have no color or odor, and are *not* oxidizing agents. They do not dissolve in TCE.

Given the solubility properties of the alkaline earth cations, and the oxidizing power of the halogens, it is possible to develop a systematic procedure for determining the presence of any Group 2 cation and any Group 7 anion in a solution. In the last part of this experiment you will be asked to set up such a procedure and use it to establish the identity of an unknown solution containing a single alkaline earth halide.

EXPERIMENTAL PROCEDURE

I.   Relative Solubilities of Some Salts of the Alkaline Earths.   Add about 1 ml (approximately 12 drops) of 0.1 M solutions of the nitrate salts of barium, calcium, magnesium, and strontium to separate small test tubes. To each tube add 1 ml of 1 M H_2SO_4 and stir with your glass stirring rod. (Rinse your stirring rod in a beaker of distilled water between tests.) Record your results in the table, noting whether a precipitate forms and any characteristics (such as color, tendency to settle out, or size of particles) that might distinguish it.

Repeat the experiment using 1 M Na_2CO_3 as the precipitating reagent instead of 1 M H_2SO_4, and record your observations. Then test for the solubilities of the oxalate salts with 0.25 M $(NH_4)_2C_2O_4$. Finally, determine the relative solubilities of the chromate salts, using 1 ml of 1 M K_2CrO_4 plus 1 ml of 1 M acetic acid as the testing reagent.

II.   Relative Oxidizing Powers of the Halogens.   In a small test tube place a few ml of bromine-saturated water and add 1 ml of trichloroethane. Stopper the test tube and shake until the bromine color is mostly in the TCE layer. (*CAUTIONS:* Avoid breathing the halogen vapors. Don't use your finger to stopper the tube, since a halogen solution can give you a bad chemical burn.) Repeat the experiment using chlorine water and iodine water with separate samples of TCE, noting any color changes as the bromine, chlorine, and iodine are extracted from the water layer into the TCE layer.

Shake 1 ml of bromine water with 1 ml of TCE in a small test tube. Add 1 ml 0.1 M NaCl solution, stopper, and shake. Using another sample of Br_2 solution with TCE, repeat the experiment using 0.1 M NaI solution. In each case observe the color of the TCE phase before and after addition of the halide. If the color of the TCE phase changes, a reaction between Br_2 and the halide ion must have occurred. Such a reaction indicates that Br_2 is a stronger oxidizing agent than the halogen formed from that halide ion. Repeat the tests on 1-ml samples of the three halide solutions, using chlorine water and then iodine water in TCE. Since mixtures of Cl^- with Cl_2 and I^- with I_2 will not aid in deciding on relative oxidizing strengths, you will not need to test them. Record your observations for each test.

III. Identification of an Alkaline Earth Halide. Your observations on the solubility properties of the alkaline earth cations should allow you to develop a method for determining which of those cations is present in a solution containing one Group 2 cation and no other cations. The method will involve testing samples of the solution with one or more of the reagents you used in Part I. Indicate on the data page how you would proceed.

In a similar way you can determine which halide ion is present in a solution containing only one such anion and no others. There you will need to test a solution of an oxidizing halogen with your unknown to see how the halide ion is affected. From the behavior of the halogen-halide ion mixtures you studied in Part II you should be able to identify easily the particular halide that is present. Describe your method on the data page, obtain an unknown solution of an alkaline earth halide, and then use your procedure to determine the cation and anion that it contains.

DATA AND OBSERVATIONS: The Alkaline Earths and the Halogens

I. Solubilities of Salts of the Alkaline Earths

	1 M H_2SO_4	1 M Na_2CO_3	0.25 M $(NH_4)_2C_2O_4$	1 M K_2CrO_4 1 M acetic acid
$Ba(NO_3)_2$				
$Ca(NO_3)_2$				
$Mg(NO_3)_2$				
$Sr(NO_3)_2$				

In filling in the table, use the following key: P = precipitate forms; S = no precipitate. If you get a precipitate, describe its color and tendency to settle out.

Consider the relative solubilities of the Group 2 cations in the various precipitating reagents. On the basis of the trends you observed, list the four alkaline earths in the order in which they should appear in the Periodic Table. *Start with the one which forms the most soluble oxalate.*

_____   _____   _____   _____

Why did you arrange the elements as you did? Is the order consistent with the properties of the cations in all of the precipitating reagents?

II. Relative Oxidizing Powers of the Halogens

a. Color of the halogen in solution:

	Br_2	Cl_2	I_2
Water	_____	_____	_____
TCE	_____	_____	_____

Continued on following page   **83**

b. Reactions between halogens and halides:

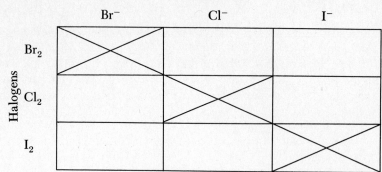

State initial and final colors of TCE layer. R = reaction occurs; NR = no reaction occurs.

III.   Identification of an Alkaline Earth Halide

Procedure for identifying the Group 2 cation:

Procedure for identifying the Group 7 anion:

Observations on unknown alkaline earth halide solution:

Cation present _____

Anion present _____

Unknown no. _____

ADVANCE STUDY ASSIGNMENT: The Alkaline Earths and the Halogens

1.   Carbon has a density of about 2.3 g/cm^3 when in the form of graphite. Germanium has a density of about 5.3 g/cm^3. Using the Periodic Table, predict whether silicon would have a density greater than that of germanium. Make the same kind of prediction for tin. Indicate your reasoning.

2.   Substances A, B, and C can all act as oxidizing agents. In the reactions in which they participate, they are reduced to A^-, B^-, and C^- ions. When a solution of A is mixed with one containing B^- ions, an oxidation-reduction reaction occurs. Write the equation for the reaction:

Which species is oxidized? _____

Which is reduced? _____

When a solution of A is mixed with one containing C^- ions, no reaction occurs.

Is A a better oxidizing agent than B? _____

Is A a better oxidizing agent than C? _____

Arrange A, B, and C in order of increasing strengths as oxidizing agents.

3.   You are given an unknown solution which may contain only one salt from the following set: NaA, NaB, NaC. In solution each salt dissociates completely into Na^+ ion and the anion A^-, B^-, or C^-, whose properties are those in Problem 2. The Na^+ ion is effectively inert. Given a way to determine whether the reactions in Problem 2 occur, develop a simple procedure for identifying the salt present in your unknown. You can assume that substances A, B, and C are available to you.

12 • The Geometrical Structure of Molecules: An Experiment Using Molecular Models

Many years ago it was observed that in many of its compounds the carbon atom formed four chemical linkages to other atoms. As early as 1870, graphic formulas of carbon compounds were drawn as shown:

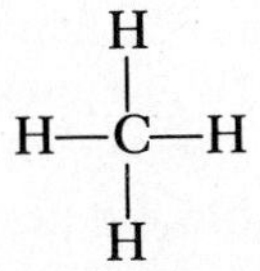

$$
\begin{array}{ccc}
& H & \\
& | & \\
H- & C & -H \\
& | & \\
& H &
\end{array}
\qquad
\begin{array}{cc}
H & H \\
| & | \\
C & = C \\
| & | \\
H & H
\end{array}
$$

methane 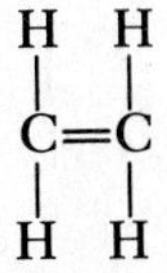   ethylene

Although such drawings as these would imply that the atom-atom linkages, indicated by valence strokes, lie in a plane, chemical evidence, particularly the existence of only one substance with the graphic formula

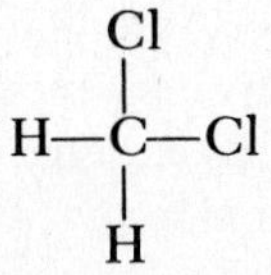

$$
\begin{array}{c}
Cl \\
| \\
H-C-Cl \\
| \\
H
\end{array}
$$

requires that the linkages be directed toward the corners of a tetrahedron, at the center of which is the carbon atom.

The physical significance of the chemical linkages between atoms, expressed by the lines or valence strokes in molecular structure diagrams, became evident soon after the discovery of the electron. In 1916 in a classic paper, G. N. Lewis suggested, on the basis of chemical evidence, that the single bonds in graphic formulas involve two electrons and that an atom tends to hold eight electrons in its outermost or valence shell.

Lewis' proposal that atoms generally have eight electrons in their outer shell proved to be extremely useful and has come to be known as the octet rule. It can be applied to many atoms, but is particularly important in the treatment of covalent compounds of atoms in the second row of the Periodic Table. For atoms such as carbon, oxygen, nitrogen, and fluorine, the eight valence electrons occur in pairs that occupy tetrahedral positions around the central atom core. Some of the electron pairs do not participate directly in chemical bonding and are called unshared or nonbonding pairs; however, the structures of compounds containing such unshared pairs reflect the tetrahedral arrangement of the four pairs of valence shell electrons. In the H_2O molecule, which obeys the octet rule, the four pairs of electrons around the central oxygen atom occupy essentially tetrahedral positions; there are two unshared nonbonding pairs and two bonding pairs which are shared by the O atom and the two H atoms. The H—O—H bond angle is nearly but not exactly tetrahedral since the properties of shared and unshared pairs of electrons are not exactly alike.

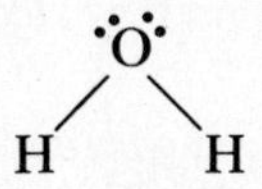

$$
\begin{array}{c}
\ddot{O} \\
/ \ \\
H \qquad H
\end{array}
$$

Most molecules obey the octet rule. Essentially, all organic molecules obey the rule, and so do most inorganic molecules and ions. For species that obey the octet rule it is possible to draw electron-dot, or Lewis, structures. The drawing of the H_2O molecule just above is an example of a Lewis structure. Below we have listed several others:

In each of these structures there are eight electrons around each atom (except for H atoms, which always have two electrons). There are two electrons in each bond. When counting electrons in these structures, one considers the electrons in a bond between two atoms as belonging to the atom under consideration. In the CH_2Cl_2 molecule just above, for example, the Cl atoms each have eight electrons, including the two in the single bond to the C atom. The C atom also has eight electrons, two from each of the four bonds to that atom. The bonding and non-bonding electrons in Lewis structures are all from the *outermost* shells of the atoms involved, and are the so-called valence electrons of those atoms. For the main group elements, the number of valence electrons in an atom is equal to the group number of the element in the Periodic Table. Carbon, in Group 4, has four valence electrons in its atoms; hydrogen, in Group 1, has one; chlorine, in Group 7, has seven valence electrons. In an octet rule structure the valence electrons from all the atoms are arranged in such a way that each atom, except hydrogen, has eight electrons.

Often it is quite easy to construct an octet rule structure for a molecule. Given that an oxygen atom has six valence electrons (Group 6) and a hydrogen atom has one, it is clear that one O and two H atoms have a total of eight valence electrons; the octet rule structure for H_2O, which we discussed earlier, follows by inspection. Structures like that of H_2O, involving only single bonds and non-bonding electron pairs are common. Sometimes, however, there is a "shortage" of electrons; that is, it is not possible to construct an octet rule structure in which all the electron pairs are either in single bonds or are nonbonding. C_2H_4 is a typical example of such a species. In such cases, octet rule structures can often be made in which two atoms are bonded by two pairs, rather than one pair, of electrons. The two pairs of electrons form a double bond. In the C_2H_4 molecule, shown above, the C atoms each get four of their electrons from the double bond. The assumption that electrons behave this way is supported by the fact that the C=C double bond is both shorter and stronger than the C—C single bond in the C_2H_6 molecule (see above). Double bonds, and triple bonds, occur in many molecules, usually between C, O, N, and/or S atoms.

Lewis structures can be used to predict molecular and ionic geometries. All that is needed is to assume that the four pairs of electrons around each atom are arranged tetrahedrally. We have seen how that assumption leads to the correct geometry for H_2O. Applying the same principle to the species whose Lewis structures we listed earlier, we would predict, correctly, that the $C_2H_2Cl_2$ molecule would be tetrahedral (roughly anyway), that NH_3 would be pyramidal (with the nonbonding electron pair sticking up from the pyramid made from the atoms, that the bond angles in C_2H_6 are all tetrahedral, and that the C_2H_4 molecule is planar (the two bonding pairs in the double bond are in a sort of banana bonding arrangement above and below the plane of the molecule. In describing molecular geometry we indicate the positions of the atomic nuclei, not the electrons. The NH_3 molecule is pyramidal, not tetrahedral.

It is also possible to predict polarity from Lewis structures. Polar molecules have their center of positive charge at a different point than their center of negative charge. This separation of charges produces a dipole moment in the molecule. Covalent bonds between different kinds of atoms are polar; all heteronuclear diatomic molecules are polar. In some molecules the polarity from one bond may be cancelled by that from others. Carbon dioxide, CO_2, which is linear, is a nonpolar molecule. Methane, CH_4, which is tetrahedral, is also nonpolar. Among the species whose Lewis structures we have listed, we find that H_2O, CH_2Cl_2, NH_3, and OH^- are polar. C_2H_6 and C_2H_4 are nonpolar.

For some molecules with a given molecular formula, it is possible to satisfy the octet rule with different atomic arrangements. A simple example would be

The two molecules are called isomers of each other, and the phenomenon is called isomerism. Although the molecular formulas of both substances are the same, C_2H_6O, their properties differ markedly because of their different atomic arrangements.

Isomerism is very common, particularly in organic chemistry, and when double bonds are present, isomerism can occur in very small molecules:

The first two isomers result from the fact that there is no rotation around a double bond, although such rotation can occur around single bonds. The third isomeric structure cannot be converted to either of the first two without breaking bonds.

With certain molecules, given a fixed atomic geometry, it is possible to satisfy the octet rule with more than one bonding arrangement. The classic example is benzene, whose molecular formula is C_6H_6:

These two structures are called resonance structures, and molecules such as benzene, which have two or more resonance structures, are said to exhibit resonance. The actual bonding in such molecules is thought to be an average of the bonding present in the resonance structures. The stability of molecules exhibiting resonance is found to be higher than that anticipated for any single resonance structure.

Although the conclusions we have drawn regarding molecular geometry and polarity can be obtained from Lewis structures, it is much easier to draw such conclusions from models of molecules and ions. The rules we have cited for octet rule structures transfer readily to models. In many ways the models are easier to construct than are the drawings of Lewis structures on paper. In addition, the models are three-dimensional and hence much more representative of the actual species. Using the models, it is relatively easy to see both geometry and polarity, as well as to deduce Lewis structures. In this experiment you will assemble models for a sizeable number of common chemical species and interpret them in the ways we have discussed.

EXPERIMENTAL PROCEDURE

In this experiment you may work in pairs during the first portion of the laboratory period.

The models you will use consist of drilled wooden balls, short sticks, and springs. The balls represent atomic nuclei surrounded by the inner electron shells. The sticks and springs represent electron pairs and fit in the holes in the wooden balls. The model (molecule or ion) consists of wooden balls (atoms) connected by sticks or springs (chemical bonds). Some sticks may be connected to only one atom (nonbonding pairs).

In this experiment we will deal with atoms that obey the octet rule; such atoms have four electron pairs around the central core and will be represented by balls with four tetrahedral holes in which there are four sticks or springs. The only exception will be hydrogen atoms, which share two electrons in covalent compounds, and which will be represented by balls with a single hole in which there is a single stick.

In assembling a molecular model of the kind we are considering, it is possible, indeed desirable, to proceed in a systematic manner. We will illustrate the recommended procedure by developing a model for a molecule with the formula CH_2O.

1. Determine the total number of valence electrons in the species. This is easily done once you realize that the number of valence electrons on an atom is equal to the number of the group to which the atom belongs in the Periodic Table. For CH_2O,

$$C \text{ Group 4} \quad H \text{ Group 1} \quad O \text{ Group 6}$$

Therefore each carbon atom in a molecule or ion contributes four electrons, each hydrogen atom one electron, and each oxygen atom six electrons. The total number of valence electrons equals the sum of the valence electrons on all of the atoms in the species being studied. For CH_2O this total would be $4 + (2 \times 1) + 6$, or 12 valence electrons. If we are working with an ion, we add one electron for each negative charge or subtract one for each positive charge on the ion.

2. Select wooden balls and sticks to represent the atoms and electron pairs in the molecule. You should use four-holed balls for the carbon atom and the oxygen atom, and one-holed balls to represent the hydrogen atoms. Since there are 12 valence electrons in the molecule and electrons occur in pairs, you will need six sticks to represent the six electron pairs. The sticks will serve both as bonds between atoms and as nonbonding electron pairs.

3. Connect the balls with some of the sticks. (Assemble a skeleton structure for the molecule, joining atoms by single bonds.) In some cases this can only be done in one way. Usually, however, there are various possibilities, some of which are more reasonable than others. In CH_2O the model can be assembled by connecting the two H atom balls to the C atom ball with two of the available sticks, and then using a third stick to connect the C atom and O atom balls.

4. The next step is to use the sticks that are left over in such a way as to fill all the remaining holes in the balls. (Distribute the electron pairs so as to give each atom eight electrons and so satisfy the octet rule.) In the model we have assembled, there is one unfilled hole in the C atom ball, three unfilled holes in the O atom ball, and three available sticks. An obvious way to meet the required condition is to use two sticks to fill two of the holes in the O atom ball, and then use two springs instead of two sticks to connect the C atom and O atom balls. The model as completed is shown in Figure 12.1.

5. Interpret the model in terms of the atoms and bonds represented. The sticks and spatial arrangement of the balls will closely correspond to the electronic and atomic arrangement in the molecule. Given our model, we would describe the CH_2O molecule as being planar with single bonds between carbon and hydrogen atoms and a double bond between the C and O atoms. The H—C—H angle is approximately tetrahedral. There are two nonbonding electron pairs on the O atom. Since all bonds are polar and the molecular symmetry does not cancel the polarity in CH_2O, the molecule is polar. The Lewis structure of the molecule is given below:

$$\overset{\cdot\cdot}{\underset{\cdot\cdot}{O}} = C \overset{\diagup H}{\diagdown_{H}}$$

(The compound having molecules with the formula CH_2O is well known and is called formaldehyde. The bonding and structure in CH_2O are as given by the model.)

6. Investigate the possibility of the existence of isomers or resonance structures. It turns out that in the case of CH_2O one can easily construct an isomeric form which obeys the octet rule, in which the central atom is oxygen rather than carbon. It is found that this isomeric form of CH_2O

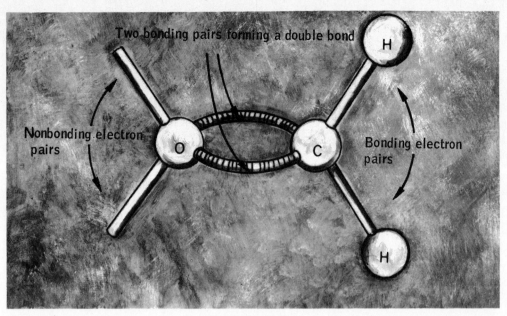

Figure 12.1

does not exist in nature. As a general rule carbon atoms almost always form a total of four bonds; put another way, nonbonding electron pairs on carbon atoms are very rare. Another useful rule of a similar nature is that if a species contains several atoms of one kind and one of another, the atoms of the same kind will assume equivalent positions in the species. In SO_4^{2-}, for example, the four O atoms are all equivalent, and are bonded to the S atom and not to each other.

Resonance structures are reasonably common. For resonance to occur, however, the atomic arrangement must remain fixed for two or more possible electronic structures. For CH_2O there are no resonance structures.

A. Using the procedure we have outlined, construct and report on models of the molecules and ions listed here and/or other species assigned by your instructor. Draw the complete Lewis structure for each molecule, showing nonbonding as well as bonding electrons.

CH_4	H_3O^+	N_2	C_2H_2
CH_2Cl_2	HF	P_4	SO_2
CH_4O	NH_3	C_2H_4	SO_4^{2-}
H_2O	H_2O_2	$C_2H_2Br_2$	CO_2

B. Assuming that stability requires that each atom obey the octet rule, predict the stability of the following species:

$$PCl_3 \qquad CH_3 \qquad OH \qquad CO$$

C. When you have completed parts A and B, see your laboratory instructor, who will check your results and assign you a set of unknown species. Working now by yourself, assemble models for each species as in the previous section, and report on the geometry and bonding in each of the unknown species on the basis of the model you construct. Also consider and report on the polarity and the likelihood of existence of isomers for each species.

EXP. 12

Name _____ Section _____

REPORT: **Geometrical Structures of Molecules Using Molecular Models**

A.

Species	Lewis Structure	Geometry	Polarity	Isomers or Resonance	Species	Lewis Structure	Geometry	Polarity	Isomers or Resonance
CH_4					NH_3				
CH_2Cl_2					H_2O_2				
CH_4O					N_2				
H_2O					P_4				
H_3O^+					C_2H_4				
HF					$C_2H_2Br_2$				

Continued on following page

Continued

94

Species	Lewis Structure	Geometry	Polarity	Isomers or Resonance	Species	Lewis Structure	Geometry	Polarity	Isomers or Resonance
C_2H_2					SO_4^{2-}				
SO_2					CO_2				

B. Stability predicted for PCl_3 _____ CH_3 _____ OH _____ CO _____

C. Unknowns _____ _____ _____   _____ _____ _____

ADVANCE STUDY ASSIGNMENT: Geometrical Structure of Molecules

You are asked by your instructor to construct a model of the CH_3Cl molecule. Being of a conservative nature, you proceed as directed in the section on Experimental Procedure.

1. First you need to find the number of valence electrons in CH_3Cl. The number of valence electrons in an atom of an element is equal to the group number of that element in the Periodic Table.

C is in Group _____   H is in Group _____   Cl is in Group _____

In CH_3Cl there is a total of _____ valence electrons.

2. The model consists of balls and sticks. What kind of ball should you select for the C atom? _____ the H atoms? _____ the Cl atom? _____ The electrons in the molecule are paired, and each stick represents an electron pair. How many sticks do you need? _____

3. Assemble a skeleton structure for the molecule, connecting the balls with sticks into one unit. Use the rule that C atoms form four bonds, whereas Cl atoms usually do not. Draw a sketch of the skeleton below:

4. How many sticks did you need to make the skeleton structure? _____ How many sticks are left over? _____ If your model is to obey the octet rule each ball must have four sticks in it (except for hydrogen atom balls, which need only one) (Each atom in an octet rule species is surrounded by 4 pairs of electrons.) How many holes remain to be filled? _____
Fill them with the remaining sticks, which represent nonbonding electron pairs. Draw the complete Lewis structure for CH_3Cl using lines for bonds and pairs of dots for nonbonding electrons.

5. Describe the geometry of the model, which is that of CH_3Cl _____ Is the CH_3Cl molecule polar? _____ Why?

Would you expect CH_3Cl to have any isomeric forms? _____ Explain your reasoning.

EXPERIMENT

13 • Vapor Pressure and Heat of Vaporization of Liquids

The vapor pressure of a pure liquid is the total pressure at equilibrium in a container in which only the liquid and its vapor are present. In a container in which the liquid and other gas are both present, the vapor pressure of the liquid is equal to the partial pressure of its vapor in the container. In this experiment you will measure the vapor pressure of a liquid by determining the increase that occurs in the pressure in a closed container filled with air when the liquid is injected into it.

The vapor pressure of a liquid rises rapidly as the temperature is increased and reaches one atmosphere at the normal boiling point of the liquid. Thermodynamic arguments show that the vapor pressure of a liquid depends on temperature according to the equation

$$\log_{10}VP = -\frac{\Delta H_{vap}}{2.3RT} + C \tag{1}$$

where VP is the vapor pressure, ΔH_{vap} is the amount of heat in joules required to vaporize one mole of the liquid against a constant pressure, R is the gas constant, 8.31 J/mole K, and T is the absolute temperature. You will note that this equation is of the form

$$Y = BX + C \tag{2}$$

where $Y = \log_{10}VP$, $X = 1/T$ and $B = -\Delta H_{vap}/2.3R$. Consequently, if we measure the vapor pressure of a liquid at various temperatures and plot $\log_{10}VP$ vs. $1/T$, we should obtain a straight line. From the slope B of this line, we can calculate the heat of vaporization of the liquid, since $\Delta H_{vap} = -2.3RB$.

In the laboratory you will measure the vapor pressure of an unknown liquid at approximately 0°C, 20°C, and 40°C, as well as its boiling point at atmospheric pressure. Given the three vapor pressures, you will be able to calculate the heat of vaporization of the liquid by making a graph of $\log_{10}VP$ vs. $1/T$. The graph will then be used to predict the boiling point of the liquid, and the value obtained will be compared with that you found experimentally.

EXPERIMENTAL PROCEDURE

WEAR YOUR SAFETY GLASSES WHILE PERFORMING THIS EXPERIMENT

From the stockroom obtain a suction flask, a rubber stopper fitted with a small dropper, and a short length of rubber tubing. Also obtain a sample of an unknown liquid.

1. Assemble the apparatus, using the mercury manometer at your lab bench, as indicated in Figure 13.1. The flask should be dry on the inside. If it is not, rinse it with a few ml of acetone (**flammable**) and blow compressed air into it for a few moments until it is dry. Be *gentle* when using compressed air. Put a piece of folded rectangular filter paper into the flask, so that liquid from the dropper will fall on the paper. This will facilitate the diffusion of the vapor. Pour some tap water into a beaker and bring it to about 20°C by adding some cold or warm water. Pour this water into the large beaker so that the level of water reaches as far as possible up the neck of the flask. Wait several minutes to ensure that the flask and air inside it are at the temperature of the water. Then remove the stopper from the flask. Pour a small amount of the unknown liquid into a small beaker, and draw about 2 ml of the liquid up into the dropper. Blot any excess liquid from

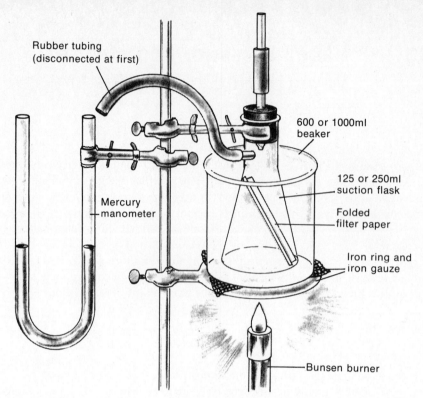

Figure 13.1

the end of the dropper with a paper towel. Press the stopper *firmly* into the flask and connect the hose to the manometer. The mercury levels in the manometer should remain essentially equal.

Immediately squeeze the liquid from the dropper onto the paper, where it will vaporize, diffuse, and exert its vapor pressure. This vapor pressure will be equal to the increase in gas pressure at equilibrium in the container. If you do not observe any appreciable (> 10 mm Hg) pressure increase within a minute or two after injecting the liquid, you probably have a leak in your apparatus and should consult your instructor. When the pressure in the flask becomes steady, in about 10 minutes, read and record the heights of the mercury levels in the manometer to ± 1 mm. An easy way to do this is to measure the height of each level above the lab bench top, using a ruler or meter stick. The *difference in height* is equal to the vapor pressure of the liquid in mm Hg at the temperature of the water bath. Record that temperature to $\pm 0.2\,°$C.

In this experiment it is essential that (1) the stopper is pressed firmly into the flask, so that the flask plus tubing plus manometer constitute a gas-tight system; (2) no liquid falls into the flask until the manometer is connected; (3) the operations of stoppering the flask, connecting the hose to the manometer, and squeezing the liquid into the flask are conducted with dispatch; (4) you take care when connecting the hose to the manometer, since mercury has a poisonous vapor and is not easily cleaned up when spilled.

Remove the flask from the system. Take out the filter paper. Dry the flask with compressed air, and dry the dropper by squeezing it several times in air.

2.   Starting again at 1, carry out the same experiment at about 0°C. Put a new piece of filter paper into the flask. Use an ice-water bath for cooling the flask. Measure the bath temperature, rather than assuming it to be 0°C.

3.   Repeat the experiment once again, this time holding the water bath at 40°C by judicious heating with a Bunsen burner. Between each run, the flask and dropper must be thoroughly dried, and the precautions noted carefully observed.

4.   In the last part of the experiment you will measure the boiling point of the liquid at the barometric pressure in the laboratory. This is done by pouring the remaining sample of unknown into a large test tube. Determine the boiling point of the liquid using the apparatus shown in

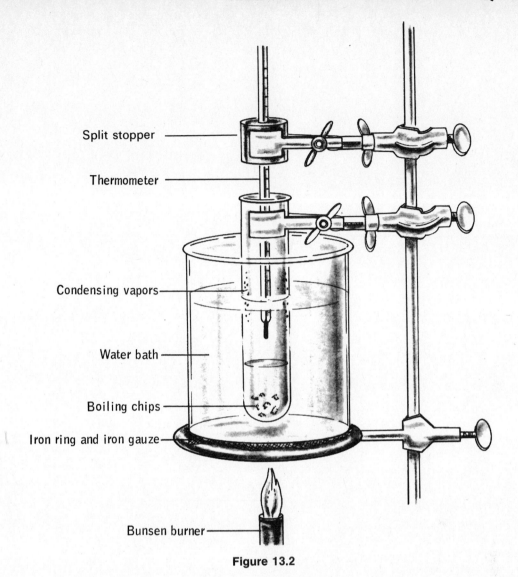

Split stopper

Thermometer

Condensing vapors

Water bath

Boiling chips

Iron ring and iron gauze

Bunsen burner

Figure 13.2

Figure 13.2. The thermometer bulb should be just above the liquid surface. Heat the water bath until the liquid in the tube boils gently, with its vapor condensing *at least 5 cm below* the top of the test tube. Boiling chips may help in keeping the liquid boiling smoothly. As the boiling proceeds there will be some condensation on the thermometer and droplets will be falling from the thermometer bulb. After a minute or two the temperature should become reasonably steady at the boiling point of the liquid. Record the temperature under these conditions, along with the barometric pressure in the laboratory. The liquids used may be flammable and toxic, so you should not inhale their vapors unnecessarily. *Do not* heat the water bath so strongly that condensation of the vapors from the liquid occurs only at the top of the test tube.

Name _____ **Section** _____

DATA AND CALCULATIONS: Vapor Pressure and Heat of Vaporization of Liquids

Temperature, t, in °C	Heights of manometer mercury levels in mm	Vapor pressure mm Hg

1. _____   _____   _____   _____

2. _____   _____   _____   _____

3. _____   _____   _____   _____

Boiling point _____ °C

Barometric pressure _____ mm Hg

Using the relation (1) between vapor pressure and temperature, we will calculate the molar heat of vaporization of your liquid and its boiling point from the vapor pressure data obtained. It would be useful to first make the calculations indicated in the following table:

Approximate temperature °C	t, actual temperature °C	T, temperature K	$1/T$	Vapor pressure, VP, in mm Hg	$\log_{10} VP$
0	_____	_____	_____	_____	_____
20	_____	_____	_____	_____	_____
40	_____	_____	_____	_____	_____

On the graph paper provided make a graph of $\log_{10} VP$ vs. $1/T$. Let $\log_{10} VP$ be the ordinate and plot $1/T$ on the abscissa. Since $\log_{10} VP$ is, by (1), a linear function of $1/T$, the line obtained should be nearly straight. Find the slope of the line, $\Delta \log_{10} VP / \Delta(1/T)$.

The slope of the line is equal to $-\Delta H_{\text{vaporization}}/2.3R$. Given that $R = 8.31$ joules/mole K, calculate the molar heat of vaporization, ΔH_{vap}, of your liquid.

$$\text{Slope} = \frac{\Delta \log_{10} VP}{\Delta 1/T} = \text{_____} = \frac{-\Delta H_{\text{vap}}}{2.3R}, \qquad \Delta H_{\text{vap}} = \text{_____} \text{joules/mole}$$

From the graph it is also possible to find the temperature at which your liquid will have any given vapor pressure. Recalling that a liquid will boil in an open container at the *temperature* at which its *vapor pressure* is equal to the *atmospheric pressure*, predict the boiling point of the liquid at the barometric pressure in the laboratory.

Continued on following page

$P_{\text{barometric}}$ ——————— mm Hg

$\log_{10} P_{\text{barometric}}$ ———————

$1/T$ at this pressure ——————— K^{-1} (from graph)

T at this pressure ——————— K

t at this pressure ——————— °C (boiling point predicted)

Boiling point observed experimentally ——————— °C

Unknown no. ———————

Does your data appear to follow the equation

$$\log_{10} VP = \frac{-\Delta H_{\text{vap}}}{2.3RT} + C$$

Give your reasoning.

**VAPOR PRESSURE AND HEAT OF
VAPORIZATION OF LIQUIDS**
(Data and Calculations)

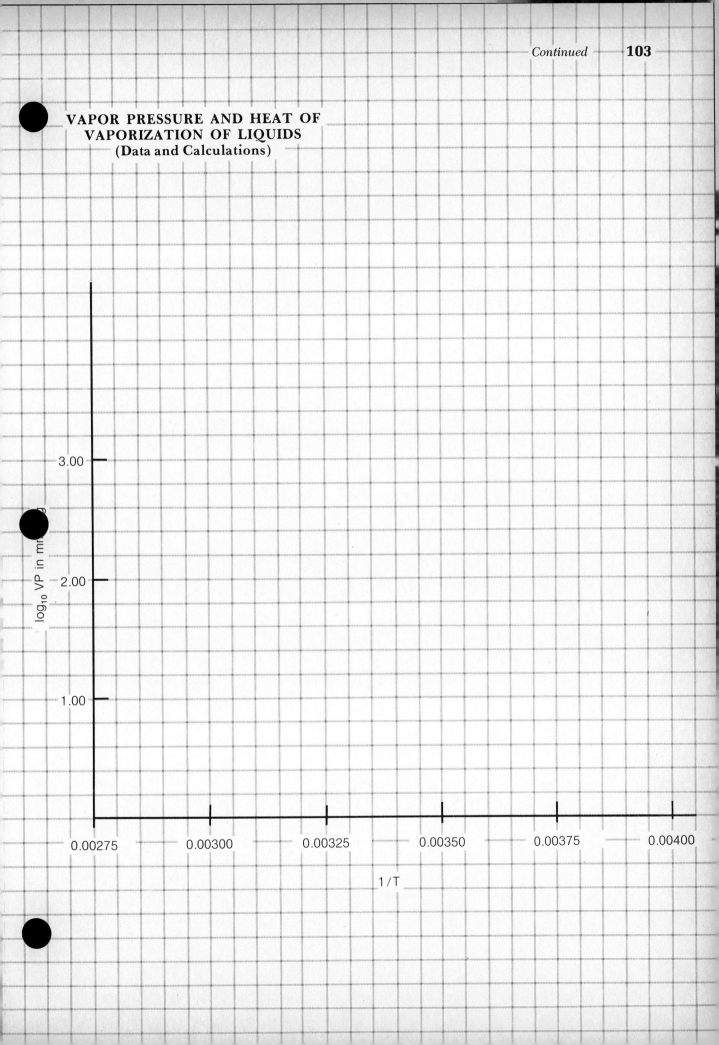

ADVANCE STUDY ASSIGNMENT: Vapor Pressure of Liquids

1. The vapor pressure of ethyl alcohol was measured by the method of this experiment with the following results (see Fig. 14.1):

$t°C$	Heights of manometer levels in mm Hg		Vapor pressure in mm Hg	$\log_{10} VP$	T in K	$1/T$
	Right	Left				
-2	84.5	94.5	_____	_____	_____	_____
19	69.5	109.5	_____	_____	_____	_____
35	39.5	139.5	_____	_____	_____	_____

a. Calculate the vapor pressure of ethyl alcohol at the three temperatures. Find the logarithm to base 10 of each vapor pressure. Enter your results in the table.

b. Find the absolute temperature T and its reciprocal at each temperature.

c. Using the graph paper on the next page, plot the three points relating $\log_{10} VP$ to $1/T$. Draw the best straight line you can through these points. (Such a line minimizes the sum of the distances from the points to the line.)

d. Determine the slope of the straight line. To do this you need to find $\Delta\log_{10} VP/\Delta(1/T)$. Perhaps the easiest way to evaluate these terms is to extend the straight line until it cuts the ordinate (the vertical axis). Take a section of the straight line, say from $1/T$ equals 0.00275 to 0.00375. For that section, $\Delta 1/T$ equals $0.00375 - 0.00275$, or 0.00100. $\Delta\log_{10} VP$ equals the change in $\log_{10} VP$ when you go from the left end of the line to the right ($\Delta\log_{10} VP$ is negative). To obtain the slope, divide $\Delta\log_{10} VP$ by $\Delta(1/T)$.

slope = _____

e. Noting that $\Delta H_{vap} = -2.3R \times$ slope, calculate the molar heat of vaporization of ethyl alcohol. (Heats of vaporization of typical liquids are about 30,000 joules/mole.)

$\Delta H_{vap} = $ _____ joules/mole

f. Ethyl alcohol will boil when its vapor pressure exceeds the air pressure at its surface. Using the graph from Part c, find $1/T$ when the vapor pressure is 760 mm Hg. Then find T in K and in °C. This temperature is the normal boiling point of ethyl alcohol.

$1/T = $ _____ ; $T = $ _____ K; $t = $ _____ °C; $t_{obs} = 78.4°C$

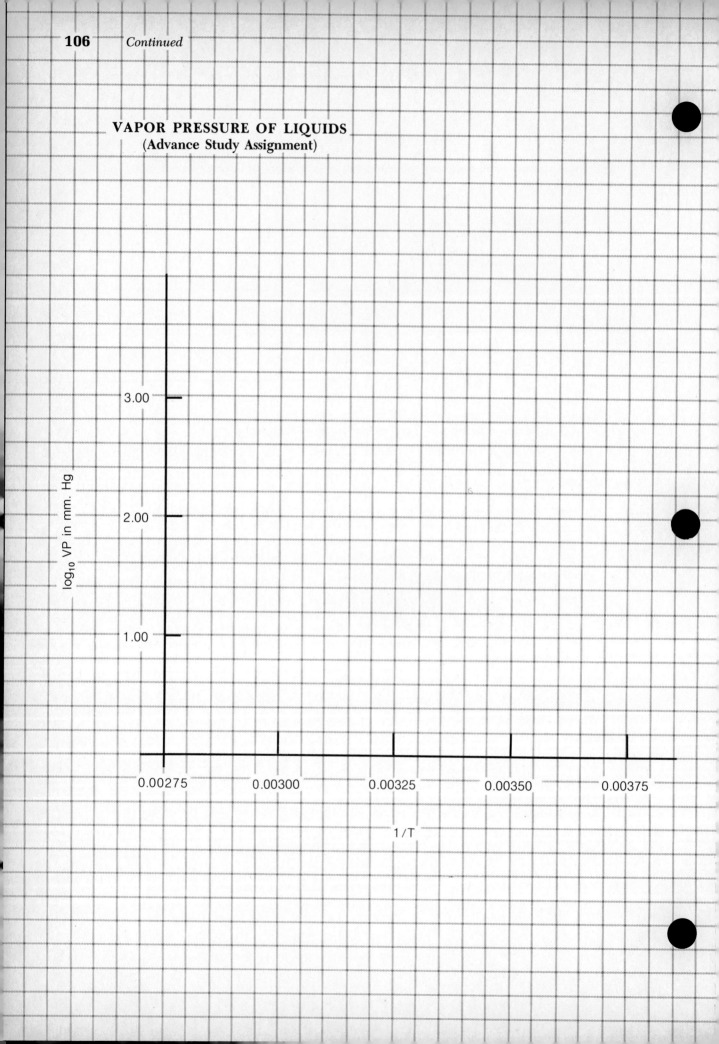

VAPOR PRESSURE OF LIQUIDS
(Advance Study Assignment)

EXPERIMENT

14 • Determination of the Equilibrium Constant for a Chemical Reaction

When chemical substances react, the reaction typically does not go to completion. Rather, the system goes to some intermediate state in which both the reactants and products have concentrations which do not change with time. Such a system is said to be in chemical equilibrium. When in equilibrium at a particular temperature, a reaction mixture obeys the Law of Chemical Equilibrium, which imposes a condition on the concentrations of reactants and products. This condition is expressed in the equilibrium constant K_c for the reaction.

In this experiment we will study the equilibrium properties of the reaction between iron(III) ion and thiocyanate ion:

$$Fe^{3+}(aq) + SCN^-(aq) \rightleftharpoons FeSCN^{2+}(aq) \tag{1}$$

When solutions containing Fe^{3+} ion and thiocyanate ion are mixed, Reaction 1 occurs to some extent, forming the $FeSCN^{2+}$ complex ion, which has a deep red color. As a result of the reaction, the equilibrium amounts of Fe^{3+} and SCN^- will be less than they would have been if no reaction occurred; for every mole of $FeSCN^{2+}$ that is formed, one mole of Fe^{3+} and one mole of SCN^- will react. According to the Law of Chemical Equilibrium, the equilibrium constant expression K_c for Reaction 1 is formulated as follows:

$$\frac{[FeSCN^{2+}]}{[Fe^{3+}][SCN^-]} = K_c \tag{2}$$

The value of K_c in Equation 2 is constant at a given temperature. This means that mixtures containing Fe^{3+} and SCN^- will react until Equation 2 is satisfied, so that the same value of the K_c will be obtained no matter what initial amounts of Fe^{3+} and SCN^- were used. Our purpose in this experiment will be to find K_c for this reaction for several mixtures made up in different ways, and to show that K_c indeed has the same value in each of the mixtures. The reaction is a particularly good one to study because K_c is of a convenient magnitude and the color of the $FeSCN^{2+}$ ion makes for an easy analysis of the equilibrium mixture.

The mixtures will be prepared by mixing solutions containing known concentrations of iron(III) nitrate, $Fe(NO_3)_3$, and potassium thiocyanate, KSCN. The color of the $FeSCN^{2+}$ ion formed will allow us to determine its equilibrium concentration. Knowing the initial composition of a mixture and the equilibrium concentration of $FeSCN^{2+}$, we can calculate the equilibrium concentrations of the rest of the pertinent species and then determine K_c.

Since the calculations required in this experiment may not be apparent, we will go through a step by step procedure by which they can be made. As a specific example, let us assume that we prepare a mixture by mixing 10.0 ml of 2.00×10^{-3} M $Fe(NO_3)_3$ with 10.0 ml of 2.00×10^{-3} M KSCN. As a result of Reaction 1, some red $FeSCN^{2+}$ ion is formed. By the method of analysis described later, its concentration at equilibrium is found to be 1.50×10^{-4} M. Our problem is to find K_c for the reaction from this information. To do this we first need to find the initial number of moles of each reactant in the mixture. Second, we determine the number of moles of product that were formed at equilibrium. Since the product was formed at the expense of reactants, we can calculate the amount of each reactant that was used up. In the third step we find the number of moles of each reactant remaining in the equilibrium mixture. Fourth, we determine the concentration of each reactant. And, finally, in the fifth step, we evaluate K_c for the reaction.

Step 1. Finding the Initial Number of Moles of Each Reactant. This requires relating the volumes and concentrations of the reagent solutions that were mixed to the numbers of moles of each reactant species in those solutions. By the definition of the molarity, M_A, of a species A,

$$M_A = \frac{\text{no. moles } A}{\text{no. liters of solution, } V} \quad \text{or} \quad \text{no. moles } A = M_A \times V \tag{3}$$

Using Equation 3, it is easy to find the initial number of moles of Fe^{3+} and SCN^-. For each solution the volume used was 10.0 ml, or 0.0100 liter. The molarity of each of the solutions was 2.00×10^{-3} M, so $M_{Fe^{3+}} = 2.00 \times 10^{-3}$ M and $M_{SCN^-} = 2.00 \times 10^{-3}$ M. Therefore, in the reagent solutions, we find that

$$\text{initial no. moles } Fe^{3+} = M_{Fe^{3+}} \times V = 2.00 \times 10^{-3} \text{ M} \times 0.0100 \text{ liter} = 20.0 \times 10^{-6} \text{ moles}$$

$$\text{initial no. moles } SCN^- = M_{SCN^-} \times V = 2.00 \times 10^{-3} \text{ M} \times 0.0100 \text{ liter} = 20.0 \times 10^{-6} \text{ moles}$$

Step 2. Finding the Number of Moles of Product Formed. Here again we can use Equation 3 to advantage. The concentration of $FeSCN^{2+}$ was found to be 1.50×10^{-4} M at equilibrium. The volume of the mixture at equilibrium is the *sum* of the two volumes that were mixed, and is 20.0 ml, or 0.0200 liter. So,

$$\text{no. moles } FeSCN^{2+} = M_{FeSCN^{2+}} \times V = 1.50 \times 10^{-4} \text{ M} \times 0.0200 \text{ liter} = 3.00 \times 10^{-6} \text{ moles}$$

The number of moles of Fe^{3+} and SCN^- that were *used up* in producing the $FeSCN^{2+}$ must also both be equal to 3.00×10^{-6} moles since, by Equation 1, it takes *one mole* Fe^{3+} *and one mole* SCN^- to make each mole of $FeSCN^{2+}$.

Step 3. Finding the Number of Moles of Each Reactant Present at Equilibrium. In Step 1 we determined that initially we had 20.0×10^{-6} moles Fe^{3+} and 20.0×10^{-6} moles SCN^- present. In Step 2 we found that in the reaction 3.00×10^{-6} moles Fe^{3+} and 3.00×10^{-6} moles SCN^- were used up. The number of moles present at equilibrium must equal the number we started with minus the number that reacted. Therefore, *at equilibrium,*

$$\text{no. moles at equilibrium} = \text{initial no. moles} - \text{no. moles used up}$$
$$\text{equil. no. moles } Fe^{3+} = 20.0 \times 10^{-6} - 3.00 \times 10^{-6} = 17.0 \times 10^{-6} \text{ moles} \tag{4}$$
$$\text{equil. no. moles } SCN^- = 20.0 \times 10^{-6} - 3.00 \times 10^{-6} = 17.0 \times 10^{-6} \text{ moles}$$

Step 4. Find the Concentrations of All Species at Equilibrium. Experimentally, we obtained the concentration of $FeSCN^{2+}$ directly. $[FeSCN^{2+}] = 1.50 \times 10^{-4}$ M. The concentrations of Fe^{3+} and SCN^- follow from Equation 3. The number of moles of each of these species at equilibrium was obtained in Step 3. The volume of the mixture being studied was 20.0 ml, or 0.0200 liter. So, *at equilibrium,*

$$[Fe^{3+}] = M_{Fe^{3+}} = \frac{\text{no. moles } Fe^{3+}}{\text{volume of solution}} = \frac{17.0 \times 10^{-6} \text{ moles}}{0.0200 \text{ liter}} = 8.5 \times 10^{-4} \text{ M}$$

$$[SCN^-] = M_{SCN^-} = \frac{\text{no. moles } SCN^-}{\text{volume of solution}} = \frac{17.0 \times 10^{-6} \text{ moles}}{0.0200 \text{ liter}} = 8.5 \times 10^{-4} \text{ M}$$

Step 5. Finding the Value of K_c for the Reaction. Once the equilibrium concentrations of all the reactants and products are known, one needs merely to substitute into Equation 2 to determine K_c:

$$K_c = \frac{[FeSCN^{2+}]}{[Fe^{3+}][SCN^-]} = \frac{1.50 \times 10^{-4}}{(8.5 \times 10^{-4}) \times (8.5 \times 10^{-4})} = 208$$

In this experiment you will obtain data similar to that shown in this example. The calculations involved in processing that data are completely analogous to those we have made. (Actually, your results will differ from the ones we obtained, since the data in our example were obtained at a different temperature and so relate to a different value of K_c.)

In carrying out this analysis we made the assumption that the reaction which occurred was given by Equation 1. There is no inherent reason why the reaction might not have been

$$Fe^{3+}(aq) + 2\ SCN^-(aq) \rightleftharpoons Fe(SCN)_2^+(aq) \tag{5}$$

If you are interested in matters of this sort, you might ask how we know whether we are actually observing Reaction 1 or Reaction 5. The line of reasoning is the following. If Reaction 1 is occurring, K_c for that reaction as we calculate it should remain constant with different reagent mixtures. If, however, Reaction 5 is the one that is going on, K_c as calculated for that reaction should remain constant. In the optional part of the Calculations section, we will assume that Reaction 5 occurs, and make the analysis of K_c on that basis. The results of the two sets of calculations should make it clear that Reaction 1 is the one that we are studying.

Two analytical methods can be used to determine $[FeSCN^{2+}]$ in the equilibrium mixtures. The more precise method uses a spectrophotometer, which measures the amount of light absorbed by the red complex at 447 nm, the wavelength at which the complex most strongly absorbs. The absorbance, A, of the complex is proportional to its concentration, M, and can be measured directly on the spectrophotometer:

$$A = kM \tag{6}$$

Your instructor will show you how to operate the spectrophotometer, if one is available to your laboratory, and will provide you with a calibration curve or equation from which you can find $[FeSCN^{2+}]$ once you have determined the absorbance of your solutions.

In the other analytical method a solution of known concentration of $FeSCN^{2+}$ is prepared. The $[FeSCN^{2+}]$ concentrations in the solutions being studied are found by comparing the color intensities of these solutions with that of the known. The method involves matching the color intensity of a given depth of unknown solution with that for an adjusted depth of known solution. The actual procedure and method of calculation are discussed in the Experimental section.

In preparing the mixtures in this experiment we will maintain the concentration of H^+ ion at 0.5 M. The hydrogen ion does not participate directly in the reaction, but its presence is necessary to avoid the formation of brown-colored species such as $FeOH^{2+}$, which would interfere with the analysis of $[FeSCN^{2+}]$.

EXPERIMENTAL PROCEDURE

Label five regular test tubes 1 to 5, with labels or by noting their positions on your test tube rack. Pour about 30 ml 2.00×10^{-3} M $Fe(NO_3)_3$ in 1 M HNO_3 into a dry 100-ml beaker. Pipet 5.0 ml of that solution into each test tube. Then add about 20 ml 2.00×10^{-3} M KSCN to another dry 100-ml beaker. Pipet 1, 2, 3, 4, and 5 ml from the KSCN beaker into each of the corresponding test tubes labeled 1 to 5. Then pipet the proper number of ml of water into each test tube to bring the total volume in each tube to 10.0 ml. The volumes of reagents to be added to each tube are summarized in Table 14.1.

TABLE 14.1

	Test Tube No.				
	1	2	3	4	5
Volume $Fe(NO_3)_3$ solution (ml)	5.0	5.0	5.0	5.0	5.0
Volume KSCN solution (ml)	1.0	2.0	3.0	4.0	5.0
Volume H_2O (ml)	4.0	3.0	2.0	1.0	0.0

Mix each solution thoroughly with a glass stirring rod. Be sure to dry the stirring rod after mixing each solution.

Method I.   Analysis by Spectrophotometric Measurement.   Place a portion of the mixture in tube 1 in a spectrophotometer cell, as demonstrated by your instructor, and measure the absorbance of the solution at 447 nm. Determine the concentration of $FeSCN^{2+}$ from the calibration curve provided for each instrument or from the equation furnished to you. Record the value on the data page. Repeat the measurement using the mixtures in each of the other test tubes.

Method II.   Analysis by Comparison with a Standard.   Prepare a solution of known $[FeSCN^{2+}]$ by pipetting 10.0 ml 0.200 M $Fe(NO_3)_3$ in 1 M HNO_3 into a test tube and adding 2.0 ml 0.002 M KSCN and 8.0 ml water. Mix the solution thoroughly with a stirring rod.

Since in this solution $[Fe^{3+}] \gg [SCN^-]$, Reaction 1 is driven strongly to the right. You can assume without serious error that essentially all the SCN^- added is converted to $FeSCN^{2+}$. Assuming that this is the case, calculate $[FeSCN^{2+}]$ in the standard solution and record the value on the data page.

The $[FeSCN^{2+}]$ in the unknown mixtures in test tubes 1 to 5 can be found by comparing the intensity of the red color in these mixtures with that of the standard solution. This can be done by placing the test tube containing mixture 1 next to a test tube containing the standard. Look down both test tubes toward a well-illuminated piece of white paper on the laboratory bench.

Pour out the standard solution into a dry clean beaker until the color intensity you see down the tube containing the standard matches that which you see when looking down the tube containing the unknown. Use a well-lit piece of white paper as your background. When the colors match, the following relation is valid:

$$[FeSCN^{2+}]_{unknown} \times \text{depth of unknown solution} = [FeSCN^{2+}] \times \text{depth of standard solution} \quad (7)$$

Measure the depths of the matching solutions with a ruler and record them. Repeat the measurement for Mixtures 2 through 5, recording the depth of each unknown and that of the standard solution which matches it in intensity.

DATA: **Determination of the Equilibrium Constant for a Chemical Reaction**

Mixture	Volume in ml, 2.00×10^{-3} M Fe(NO$_3$)$_3$	Volume in ml, 2.00×10^{-3} M KSCN	Volume in ml of Water	Method I Absorbance	Method II Depth in mm Standard	Method II Depth in mm Unknown	[FeSCN^{2+}]
1	5.0	1.0	4.0	——	——	——	—— $\times 10^{-4}$ M
2	5.0	2.0	3.0	——	——	——	—— $\times 10^{-4}$ M
3	5.0	3.0	2.0	——	——	——	—— $\times 10^{-4}$ M
4	5.0	4.0	1.0	——	——	——	—— $\times 10^{-4}$ M
5	5.0	5.0	0.0	——	——	——	—— $\times 10^{-4}$ M

If Method II was used, [FeSCN^{2+}]$_{standard}$ = ——— $\times 10^{-4}$ M; [FeSCN^{2+}] in Mixtures 1 to 5 found by Equation 7.

CALCULATIONS

A. Calculation of K_c assuming the reaction $Fe^{3+}(aq) + SCN^-(aq) \rightleftharpoons FeSCN^{2+}(aq)$.   (1) This calculation is most easily done by Steps 1 through 5 in the discussion. Results are to be entered in the table on the following page.

Step 1. Find the initial number of moles of Fe^{3+} and SCN^- in the mixtures in test tubes 1 through 5. Use Equation 3 and enter the values in the first two columns in the table.

Step 2. Enter the experimentally determined value of [FeSCN^{2+}] at equilibrium for each of the mixtures in the next to last column in the table. Use Equation 3 to find the number of moles of FeSCN^{2+} in each of the mixtures, and enter the values in the fifth column in the table. Note that this is also the number of moles of Fe^{3+} and SCN^- that were used up in the reaction.

112   *Continued*

Step 3. From the number of moles of Fe^{3+} and SCN^- initially present in each mixture, and the number of moles of Fe^{3+} and SCN^- used up in forming $FeSCN^{2+}$, calculate the number of moles of Fe^{3+} and SCN^- that remain in each mixture at equilibrium. Use Equation 4. Enter the results in Columns 3 and 4 in the table.

Step 4. Use Equation 3 and the results of Step 3 to find the concentrations of all of the species at equilibrium. The volume of the mixture is 10.0 ml, or 0.010 liter in all cases. Enter the values in Columns 6 and 7 in the table.

Step 5. Calculate K_c for the reaction for each of the mixtures by substituting values for the equilibrium concentrations of Fe^{3+}, SCN^-, and $FeSCN^{2+}$ in Equation 2.

Mixture	Initial No. Moles		Equilibrium No. Moles			Equilibrium Concentrations			K_c
	Fe^{3+}	SCN^-	Fe^{3+}	SCN^-	$FeSCN^{2+}$	$[Fe^{3+}]$	$[SCN^-]$	$[FeSCN^{2+}]$	
1	___ × 10⁻⁶	___ × 10⁻⁶	___ × 10⁻⁶	___ × 10⁻⁶	___ × 10⁻⁶	___ × 10⁻⁴	___ × 10⁻⁴	___ × 10⁻⁴	___
2	___	___	___	___	___	___	___	___	___
3	___	___	___	___	___	___	___	___	___
4	___	___	___	___	___	___	___	___	___
5	___	___	___	___	___	___	___	___	___

Continued on following page

B. **(Optional)** In calculating K_c in Part A, we assume, correctly, that the formula of the complex ion is $FeSCN^{2+}$. It is by no means obvious that this is the case and one might have assumed, for instance, that $Fe(SCN)_2^+$ was the species formed. The reaction would then be

$$Fe^{3+}(aq) + 2SCN^-(aq) \rightleftharpoons Fe(SCN)_2^+(aq) \tag{5}$$

If we analyze the equilibrium system we have studied, assuming that Reaction 5 occurs rather than Reaction 1, we would presumably obtain nonconstant values of K_c. Using the same kind of procedure as in Part A, calculate K_c for Mixtures 1, 3, and 5 on the basis that $Fe(SCN)_2^+$ is the formula of the complex ion formed by the reaction between Fe^{3+} and SCN^-. Because of the procedure used for calibrating the system by Method I or Method II, $[Fe(SCN)_2^+]$ will equal *one-half* the $[FeSCN^{2+}]$ obtained for each solution in Part A. Note that *two* moles SCN^- are needed to form *one* mole $Fe(SCN)_2^+$. This changes the expression for K_c. Also, in calculating the equilibrium number of moles SCN^- you will need to subtract ($2 \times$ number of moles $Fe(SCN)_2^+$) from the initial number of moles SCN^-.

Mixture	Initial No. Moles		Equilibrium No. Moles			Equilibrium Concentrations			K_c
	Fe^{3+}	SCN^-	Fe^{3+}	SCN^-	$Fe(SCN)_2^+$	$[Fe^{3+}]$	$[SCN^-]$	$[Fe(SCN)_2^+]$	
1	___	___	___	___	___	___	___	___	___
3	___	___	___	___	___	___	___	___	___
5	___	___	___	___	___	___	___	___	___

On the basis of the results of Part A, what can you conclude about the validity of the equilibrium concept, as exemplified by Equation 2? What do you conclude about the formula of the iron(III) thiocyanate complex ion?

Name _____ Section _____

ADVANCE STUDY ASSIGNMENT: Determination of the Equilibrium Constant
                                for a Reaction

1.   A student mixes 5.0 ml 2.00×10^{-3} M $Fe(NO_3)_3$ with 5.0 ml 2.00×10^{-3} M KSCN. She finds that in the equilibrium mixture the concentration of $FeSCN^{2+}$ is 1.4×10^{-4} M. Find K_c for the reaction $Fe^{3+}(aq) + SCN^-(aq) \rightleftharpoons FeSCN^{2+}(aq)$.

Step 1.   Find the number of moles Fe^{3+} and SCN^- initially present. (Use Eq. 3).

_____ moles Fe^{3+}; _____ moles SCN^-

Step 2.   How many moles $FeSCN^{2+}$ are in the mixture at equilibrium? What is the volume of the equilibrium mixture? (Use Eq. 3.)

_____ ml; _____ moles $FeSCN^{2+}$

How many moles of Fe^{3+} and SCN^- are used up in making the $FeSCN^{2+}$?

_____ moles Fe^{3+}; _____ moles SCN^-

Step 3.   How many moles of Fe^{3+} and SCN^- remain in the solution at equilibrium? (Use Eq. 4 and the results of Steps 1 and 2.)

_____ moles Fe^{3+}; _____ moles SCN^-

Step 4.   What are the concentrations of Fe^{3+}, SCN^-, and $FeSCN^{2+}$ at equilibrium? What is the volume of the equilibrium mixture? (Use Eq. 3 and the results of Step 3.)

$[Fe^{3+}] =$ _____ M; $[SCN^-] =$ _____ M; $[FeSCN^{2+}] =$ _____ M

_____ ml

Step 5.   What is the value of K_c for the reaction? (Use Eq. 2 and the results of Step 4.)

$K_c =$ _____

2.   **(Optional)** Assume that the reaction studied in Problem 1 is $Fe^{3+}(aq) + 2 \ SCN^-(aq) \rightleftharpoons Fe(SCN)_2^+(aq)$. Find K_c for this reaction, given the data in Problem 1, except that the equilibrium concentration of $Fe(SCN)_2^+$ is equal to 0.7×10^{-4} M.
   a. Formulate the expression for K_c for the alternate reaction just cited.

Continued on following page   **115**

b. Find K_c as you did in Problem 1; take due account of the fact that two moles SCN^- are used up per mole $Fe(SCN)_2^+$ formed.

Step 1.   Results are as in Problem 1.

Step 2.   How many moles of $Fe(SCN)_2^+$ are in the mixture at equilibrium? (Use Eq. 3.)

_____ moles $Fe(SCN)_2^+$

How many moles of Fe^{3+} and SCN^- are used up in making the $Fe(SCN)_2^+$?

_____ moles Fe^{3+}; _____ moles SCN^-

Step 3.   How many moles of Fe^{3+} and SCN^- remain in solution at equilibrium? Use the results of Steps 1 and 2, noting that no. moles SCN^- at equilibrium = original no. moles SCN^- − $(2 \times$ no. moles $Fe(SCN)_2^+)$.

_____ moles Fe^{3+}; _____ moles SCN^-

Step 4.   What are the concentrations of Fe^{3+}, SCN^-, and $Fe(SCN)_2^+$ at equilibrium? (Use Eq. 3 and the results of Step 3.)

$[Fe^{3+}] =$ _____ M; $[SCN^-] =$ _____ M; $[Fe(SCN)_2^+] =$ _____ M

Step 5.   Calculate K_c on the basis that the alternate reaction occurs. (Use the answer to Part a.)

$K_c =$ _____

EXPERIMENT

15 • Rates of Chemical Reactions, I. The Iodination of Acetone

The rate at which a chemical reaction occurs depends on several factors: the nature of the reaction, the concentrations of the reactants, the temperature, and the presence of possible catalysts. All of these factors can markedly influence the observed rate of reaction.

Some reactions at a given temperature are very slow indeed; the oxidation of gaseous hydrogen or wood at room temperature would not appreciably proceed in a century. Other reactions are essentially instantaneous; the precipitation of silver chloride when solutions containing silver ions and chloride ions are mixed and the formation of water when acidic and basic solutions are mixed are examples of extremely rapid reactions. In this experiment we will study a reaction which, in the vicinity of room temperature, proceeds at a moderate, relatively easily measured rate.

For a given reaction, the rate typically increases with an increase in the concentration of any reactant. The relation between rate and concentration is a remarkably simple one in many cases, and for the reaction

$$aA + bB \rightarrow cC$$

the rate can usually be expressed by the equation

$$\text{rate} = k(A)^m (B)^n \tag{1}$$

where m and n are generally, but not always, integers, 0, 1, 2, or possibly 3; (A) and (B) are the concentrations of A and B (ordinarily in moles per liter); and k is a constant, called the *rate constant* of the reaction, which makes the relation quantitatively correct. The numbers m and n are called the *orders of the reaction* with respect to A and B. If m is 1 the reaction is said to be *first order* with respect to the reactant A. If n is 2 the reaction is *second order* with respect to reactant B. The overall order is the sum of m and n. In this example the reaction would be *third order* overall.

The rate of a reaction is also significantly dependent on the temperature at which the reaction occurs. An increase in temperature increases the rate, an often cited rule being that a 10°C rise in temperature will double the rate. This rule is only approximately correct; nevertheless, it is clear that a rise of temperature of say 100°C could change the rate of a reaction appreciably.

As with the concentration, there is a quantitative relation between reaction rate and temperature, but here the relation is somewhat more complicated. This relation is based on the idea that in order to react, the reactant species must have a certain minimum amount of energy present at the time the reactants collide in the reaction step; this amount of energy, which is typically furnished by the kinetic energy of motion of the species present, is called the *activation energy* for the reaction. The equation relating the rate constant k to the absolute temperature T and the activation energy E_a is

$$\log_{10} k = \frac{-E_a}{2.30RT} + \text{constant} \tag{2}$$

where R is the gas constant (8.31 joules/mole K for E_a in joules per mole). This equation is identical in form to Equation 1 in Exp. 14. By measuring k at different temperatures we can determine graphically the activation energy for a reaction.

In this experiment we will study the kinetics of the reaction between iodine and acetone:

$$CH_3-\overset{\overset{\displaystyle O}{\|}}{C}-CH_3(aq) + I_2(aq) \rightarrow CH_3-\overset{\overset{\displaystyle O}{\|}}{C}-CH_2I(aq) + H^+(aq) + I^-(aq)$$

The rate of this reaction is found to depend on the concentration of hydrogen ion in the solution as well as presumably on the concentrations of the two reactants. By Equation 1, the rate law for this reaction is

$$\text{rate} = k(\text{acetone})^m (I_2)^n (H^+)^p \tag{3}$$

where m, n, and p are the orders of the reaction with respect to acetone, iodine, and hydrogen ion respectively, and k is the rate constant for the reaction.

The rate of this reaction can be expressed as the (small) change in the concentration of I_2, $\Delta(I_2)$, which occurs, divided by the time interval Δt required for the change:

$$\text{rate} = \frac{-\Delta(I_2)}{\Delta t} \tag{4}$$

The minus sign is to make the rate positive ($\Delta(I_2)$ is negative). Ordinarily, since rate varies as the concentrations of the reactants according to Equation 3, in a rate study it would be necessary to measure, directly or indirectly, the concentration of each reactant as a function of time; the rate would typically vary markedly with time, decreasing to very low values as the concentration of at least one reactant becomes very low. This makes reaction rate studies relatively difficult to carry out and introduces mathematical complexities that are difficult for beginning students to understand.

The iodination of acetone is a rather atypical reaction, in that it can be very easily investigated experimentally. First of all, iodine has color, so that one can readily follow changes in iodine concentration visually. A second and very important characteristic of this reaction is that it turns out to be zero order in I_2 concentration. This means (see Equation 3) that the rate of the reaction does not depend on (I_2) at all; $(I_2)^0 = 1$, no matter what the value of (I_2) is, as long as it is not itself zero.

Since the rate of the reaction does not depend on (I_2), we can study the rate by simply making I_2 the limiting reagent present in a large excess of acetone and H^+ ion. We then measure the time required for a known initial concentration of I_2 to be completely used up. If both acetone and H^+ are present at much higher concentrations than that of I_2, their concentrations will not change appreciably during the course of the reaction, and the rate will remain, by Equation 3, effectively constant until all the iodine is gone, at which time the reaction will stop. Under such circumstances, if it takes t seconds for the color of a solution having an initial concentration of I_2 equal to $(I_2)_0$ to disappear, the rate of the reaction, by Equation 4, would be

$$\text{rate} = \frac{-\Delta(I_2)}{\Delta t} = \frac{(I_2)_0}{t} \tag{5}$$

Although the rate of the reaction is constant during its course under the conditions we have set up, we can vary it by changing the initial concentrations of acetone and H^+ ion. If, for example, we should *double* the initial concentration of *acetone* over that in Mixture 1, keeping (H^+) and (I_2) at the *same* values they had previously, then the rate of Mixture 2 would, according to Equation 3, be different from that in Mixture 1:

$$\text{rate } 2 = k(2A)^m (I_2)^0 (H^+)^p \tag{6a}$$
$$\text{rate } 1 = k(A)^m (I_2)^0 (H^+)^p \tag{6b}$$

Dividing the first equation by the second, we see that the k's cancel, as do the terms in the iodine and hydrogen ion concentrations, since they have the same values in both reactions, and we obtain simply

$$\frac{\text{rate 2}}{\text{rate 1}} = \frac{(2A)^m}{(A)^m} = \left(\frac{2A}{A}\right)^m = 2^m \tag{6}$$

Having measured both rate 2 and rate 1 by Equation 5, we can find their ratio, which must be equal to 2^m. We can then solve for m either by inspection or using logarithms and so find the *order* of the reaction with respect to acetone.

By a similar procedure we can measure the order of the reaction with respect to H^+ ion concentration and also confirm the fact that the reaction is zero order with respect to I_2. Having found the order with respect to each reactant, we can then evaluate k, the rate constant for the reaction.

The determination of the orders m and p, the confirmation of the fact that n, the order with respect to I_2, equals zero, and the evaluation of the rate constant k for the reaction at room temperature comprise your assignment in this experiment. You will be furnished with standard solutions of acetone, iodine, and hydrogen ion, and with the composition of one solution that will give a reasonable rate. The rest of the planning and the execution of the experiment will be your responsibility.

An optional part of the experiment is to study the rate of this reaction at different temperatures in order to find its activation energy. The general procedure here would be to study the rate of reaction in one of the mixtures at room temperature and at two other temperatures, one above and one below room temperature. Knowing the rates, and hence the k's, at the three temperatures, you can then find E_a, the energy of activation for the reaction, by plotting $\log k$ vs. $1/T$. The slope of the resultant straight line, by Equation 2, must be $-E_a/2.30R$.

EXPERIMENTAL PROCEDURE

Select two regular test tubes; when filled with distilled water, they should appear to have identical color when you view them down the tubes against a white background.

Draw 50 ml of each of the following solutions into clean dry 100 ml beakers, one solution to a beaker: 4 M acetone, 1 M HCl, and 0.005 M I_2. Cover each beaker with a watch glass.

With your graduated cylinder, measure out 10.0 ml of the 4 M acetone solution and pour it into a clean 125 ml Erlenmeyer flask. Then measure out 10.0 ml 1 M HCl and add that to the acetone in the flask. Add 20.0 ml distilled H_2O to the flask. Drain the graduated cylinder, shaking out any excess water, and then use the cylinder to measure out 10.0 ml 0.005 M I_2 solution. Be careful not to spill the iodine solution on your hands or clothes.

Noting the time on your wrist watch or the wall clock to one second, pour the iodine solution into the Erlenmeyer flask and quickly swirl the flask to thoroughly mix the reagents. The reaction mixture will appear yellow because of the presence of the iodine, and the color will fade slowly as the iodine reacts with the acetone. Fill one of the test tubes $\frac{3}{4}$ full with the reaction mixture, and fill the other test tube to the same depth with distilled water. Look down the test tubes toward a well-lit piece of white paper, and note the time the color of the iodine just disappears. Measure the temperature of the mixture in the test tube.

Repeat the experiment, using as a reference the reacted solution instead of distilled water. The amount of time required in the two runs should agree within about 20 seconds.

The rate of the reaction equals the initial concentration of I_2 *in the reaction mixture* divided by the elapsed time. Since the reaction is zero order in I_2, and since both acetone and H^+ ion are present in great excess, the rate is constant throughout the reaction and the concentrations of both acetone and H^+ remain essentially at their initial values in the reaction mixture.

Having found the reaction rate for one composition of the system, it might be well to think for a moment about what changes in composition you might make to decrease the time and hence increase the rate of reaction. In particular, how could you change the composition in such a way as to allow you to determine how the rate depends upon acetone concentration? If it is not clear how to proceed, reread the discussion preceding Equation 6. In your new mixture you should keep the total volume at 50 ml, and be sure that the concentrations of H^+ and I_2 are the *same* as in the first experiment. Carry out the reaction twice with your new mixture; the times should not differ by more than about 15 seconds. The temperature should be kept within about a degree of that in the initial run. Calculate the rate of the reaction. Compare it with that for the first mixture, and then calculate the order of the reaction with respect to acetone, using a relation similar to Equation 6. First, write an equation like 6a for the second reaction mixture, substituting in the values for the rate as obtained by Equation 5 and the initial concentration of acetone, I_2, and H^+ in the reaction mixture. Then write an equation like 6b for the first reaction mixture, using the observed rate and the initial concentrations in that mixture. Obtain an equation like 6 by dividing Equation 6a by Equation 6b. Solve Equation 6 for the order m of the reaction with respect to acetone.

Again change the composition of the reaction mixture so that this time a measurement of the reaction will give you information about the order of the reaction with respect to H^+. Repeat the experiment with this mixture to establish the time of reaction to within 15 seconds, again making sure that the temperature is within about a degree of that observed previously. From the rate you determine for this mixture find p, the order of the reaction with respect to H^+.

Finally, change the reaction mixture composition in such a way as to allow you to show that the order of the reaction with respect to I_2 is zero. Measure the rate of the reaction twice, and calculate n, the order with respect to I_2.

Having found the order of the reaction for each species on which the rate depends, evaluate k, the rate constant for the reaction, from the rate and concentration data in each of the mixtures you studied. If the temperatures at which the reactions were run are all equal to within a degree or two, k should be about the same for each mixture.

Prediction of Reaction Rate (Optional)   As a final reaction, make up a mixture using reactant volumes that you did not use in any previous experiments. Using Equation 3, along with the values of the concentrations of each reactant in the mixture and the orders and the rate constant you have determined from your data, predict how long it will take for the I_2 color to disappear from your mixture. Measure the time the reaction actually takes and compare it with your predicted value.

Determination of Energy of Activation (Optional)   Select one of the reaction mixtures you have already used that gave a convenient time. Measure the rate of reaction for that mixture at about 40°C and at about 10°C. From the two rates you find, plus the rate you observed at room temperature, calculate the energy of activation for the reaction, using Equation 2.

DATA AND CALCULATIONS: The Iodination of Acetone

I. Reaction Rate Data

Mixture	Volume in ml 4 M acetone	Volume in ml 1 M HCl	Volume in ml 0.005 M I_2	Volume in ml H_2O	Time for Reaction in sec		Temp °C
					1st Run	2nd Run	
I	10	10	10	20			
II							
III							
IV							

II. Determination of Reaction Orders with Respect to Acetone, H^+ Ion, and I_2

$$\text{rate} = k \, (\text{acetone})^m \, (I_2)^n \, (H^+)^p \tag{3}$$

Calculate the *initial* concentrations of acetone, H^+ ion, and I_2 in each of the mixtures you studied. Use Equation 5 to find the rate of each reaction.

Mixture	(acetone)	(H^+)	$(I_2)_0$	$\text{rate} = \dfrac{(I_2)_0}{\text{ave. time}}$
I	0.8 M	0.2 M	0.001 M	
II				
III				
IV				

Substituting the initial concentrations and the rate from the table above, write Equation 3 as it would apply to Reaction Mixture II:

Rate II =

Now write Equation 3 for Reaction Mixture I, substituting concentrations and the calculated rate from the table:

Rate I =

Divide the equation for Mixture II by the equation for Mixture I; the resulting equation should have the ratio of Rate II to Rate I on the left side, and a ratio of acetone concentrations

Continued on following page   **121**

raised to the power m on the right. It should be similar in appearance to Equation 6. Put the resulting equation below:

$$\frac{\text{rate II}}{\text{rate I}} =$$

The only unknown in the equation is m. Solve for m. $m = \underline{\hspace{1.5cm}}$

Now write Equation 3 as it would apply to Reaction Mixture III and as it would apply to Reaction Mixture IV:

rate III =

rate IV =

Using the ratios of the rates of Mixtures III and IV to those of Mixtures II or I, find the orders of the reaction with respect to H^+ ion and I_2:

$$\frac{\text{rate III}}{\text{rate}\underline{\hspace{1cm}}} =$$ $p = \underline{\hspace{1.5cm}}$

$$\frac{\text{rate IV}}{\text{rate}\underline{\hspace{1cm}}} =$$ $n = \underline{\hspace{1.5cm}}$

III. Determination of the Rate Constant k

Given the values of m, p, and n as determined in Part II, calculate the rate constant k for each mixture by simply substituting those orders, the initial concentrations, and the observed rate from the table into Equation 3.

Mixture	I	II	III	IV	average
k	_____	_____	_____	_____	_____

IV. (Optional) Prediction of Reaction Rate

Reaction mixture

Volume in ml   Volume in ml   Volume in ml   Volume in ml
4 M acetone _____   1 M HCl _____   0.005 M I_2 _____ H_2O _____

Initial concentrations:

(acetone) _____ M (H^+) _____ M $(I_2)_0$ _____ M

Predicted rate _____ (Equation 3)

Predicted time for reaction _____ sec (Equation 5)

Observed time for reaction _____ sec

Continued on following page

V. Determination of Energy of Activation (Optional)

Reaction mixture used _____(same for all temperatures)

Time for reaction at about 10°C   _____ sec   temperature _____ °C

Time for reaction at about 40°C   _____ sec   temperature _____ °C

Time for reaction at room temp.   _____ sec   temperature _____ °C

   Calculate the rate constant at each temperature from your data, following the procedure in III.

	rate	k	log k	$\frac{1}{T(K)}$
~10°C				
~40°C				
room temp.				

   Plot log k vs. $1/T$. Find the slope of the best straight line through the points.

Slope = _____

By Equation 2:

$$E_a = -2.30 \times 8.31 \times \text{slope}$$

$$E_a = \text{_____ joules}$$

THE IODINATION OF ACETONE
(Data and Calculations)

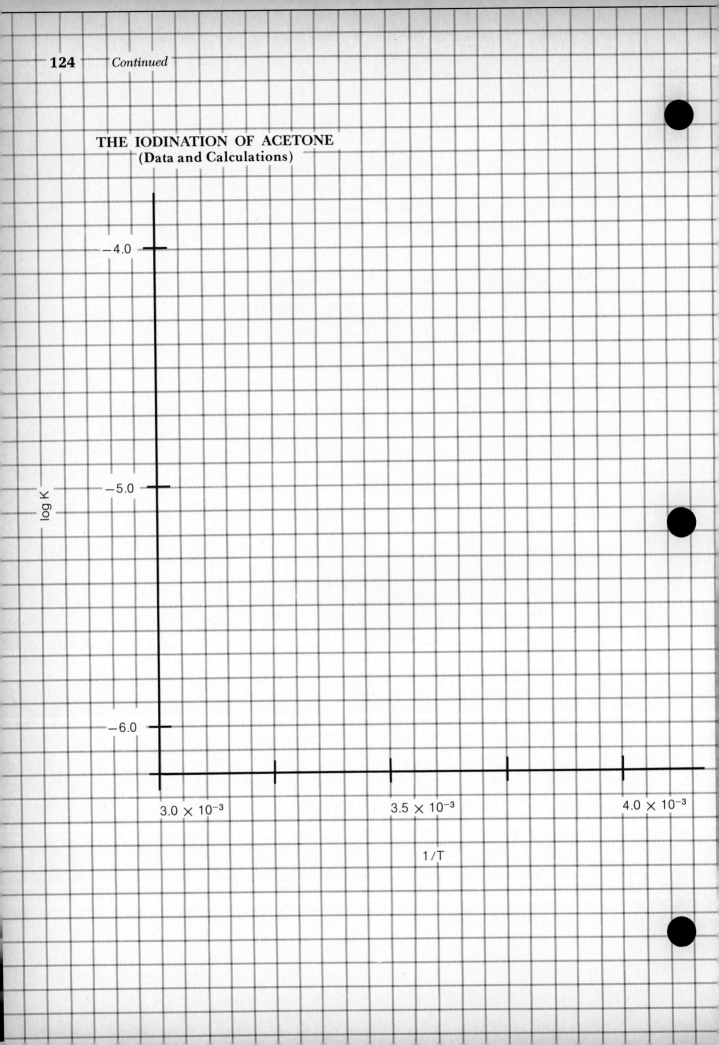

ADVANCE STUDY ASSIGNMENT: The Iodination of Acetone

1. In a reaction involving the iodination of acetone, the following volumes were used to make up the reaction mixture:

     10 ml 4 M acetone + 10 ml 1 M HCl + 10 ml 0.005 M I_2 + 20 ml H_2O

   a. How many moles of acetone were in the reaction mixture? Recall that, for a component A, no. moles $A = M_A \times V$, where M_A is the molarity of A and V the volume in liters of the solution of A that was used.

                                                ———————— moles acetone

   b. What was the molarity of acetone in the *reaction mixture?* The volume of the *mixture* was 50 ml, 0.050 liter, and the number of moles of acetone was found in Part a. Again, $M_A = \dfrac{\text{no. moles } A}{V \text{ of soln. in liters}}$

                                                ———————— M acetone

   c. How could you double the molarity of the acetone in the reaction mixture, keeping the total volume at 50 ml and keeping the same concentrations of H^+ ion and I_2 as in the original mixture?

2. Using the reaction mixture in Problem 1, a student found that it took 300 seconds for the color of the I_2 to disappear.
   a. What was the rate of the reaction? Hint: First find the initial concentration of I_2 in the reaction mixture, $(I_2)_0$. Then use Equation 5.

                                             rate = ————————

   b. Given the rate from Part a, and the initial concentrations of acetone, H^+ ion, and I_2 in the reaction mixture, write Equation 3 as it would apply to the mixture.

   rate =

Continued on following page   **125**

c. What are the unknowns that remain in the equation in Part b?

_____ _____ _____ _____

3. A second reaction mixture was made up in the following way:

    20 ml 4 M acetone + 10 ml 1 M HCl + 10 ml 0.005 M I_2 + 10 ml H_2O

a. What were the initial concentrations of acetone, H^+ ion, and I_2 in the reaction mixture?

    (acetone) _____ M; (H^+) _____ M; $(I_2)_0$ _____ M

b. It took 140 seconds for the I_2 color to disappear from the reaction mixture when it occurred at the same temperature as the reaction in Problem 2.
   What was the rate of the reaction?   _____

   Write Equation 3 as it would apply to the second reaction mixture:

   rate =

c. Divide the equation in Problem 3b by the equation in Problem 2b. The resulting equation should have the ratio of the two rates on the left side and a ratio of acetone concentrations raised to the m power on the right. Write the resulting equation and solve for the value of m, the order of the reaction with respect to acetone. (Round off the value of m to the nearest integer.)

$m =$ _____

16 • Rates of Chemical Reactions, II. A Clock Reaction

In the previous experiment we discussed the factors that influence the rate of a chemical reaction and presented the terminology used in quantitative relations in studies of the kinetics of chemical reactions. That material is also pertinent to this experiment and should be studied before you proceed further.

This experiment involves the study of the rate properties, or chemical kinetics, of the following reaction between iodide ion and bromate ion under acidic conditions:

$$6\,I^-(aq) + BrO_3^-(aq) + 6\,H^+(aq) \rightarrow 3\,I_2(aq) + Br^-(aq) + 3\,H_2O \tag{1}$$

This reaction proceeds reasonably slowly at room temperature, its rate depending on the concentrations of the I^-, BrO_3^-, and H^+ ions according to the rate law discussed in the previous experiment. For this reaction the rate law takes the form

$$\text{rate} = k(I^-)^m(BrO_3^-)^n(H^+)^p \tag{2}$$

One of the main purposes of the experiment will be to evaluate the rate constant k and the reaction orders m, n, and p for this reaction. We will also investigate the manner in which the reaction rate depends on temperature and will evaluate the activation energy E_a for the reaction.

Our method for measuring the rate of the reaction involves what is frequently called a "clock" reaction. In addition to Reaction 1, whose kinetics we will study, the following reaction will also be made to occur simultaneously in the reaction flask:

$$I_2(aq) + 2\,S_2O_3^{2-}(aq) \rightarrow 2\,I^-(aq) + S_4O_6^{2-}(aq) \tag{3}$$

As compared with (1) this reaction is essentially instantaneous. The I_2 produced in (1) reacts completely with the thiosulfate, $S_2O_3^{2-}$, ion present in the solution, so that until all the thiosulfate ion has reacted, the concentration of I_2 is effectively zero. As soon as the $S_2O_3^{2-}$ is gone from the system, the I_2 produced by (1) remains in the solution and its concentration begins to increase. The presence of I_2 is made strikingly apparent by a starch indicator which is added to the reaction mixture, since I_2 even in small concentrations reacts with starch solution to produce a blue color.

By carrying out Reaction 1 in the presence of $S_2O_3^{2-}$ and a starch indicator, we introduce a "clock" into the system. Our clock tells us when a given amount of BrO_3^- ion has reacted ($1/6$ mole BrO_3^- per mole $S_2O_3^{2-}$), which is just what we need to know, since the rate of reaction can be expressed in terms of the time it takes for a particular amount of BrO_3^- to be used up. In all our reactions, the amount of BrO_3^- that reacts in the time we measure will be constant and small as compared to the amounts of any of the other reactants. This means that the concentrations of all reactants will be essentially constant in Equation 2, and hence so will the rate during each reaction.

In our experiment we will carry out the reaction between BrO_3^-, I^-, and H^+ ions under different concentration conditions. Measured amounts of each of these ions in water solution will be mixed in the presence of a constant small amount of $S_2O_3^{2-}$. The time it takes for each mixture to turn blue will be measured. The time obtained for each reaction will be inversely proportional to its rate. By changing the concentration of one reactant and keeping the other concentrations constant, we can investigate how the rate of the reaction varies with the concentration of a particular reactant. Once we know the order for each reactant we can determine the rate constant for the reaction.

In the last part of the experiment we will investigate how the rate of the reaction depends on temperature. You will recall that in general the rate increases sharply with temperature. By measuring how the rate varies with temperature we can determine the activation energy, E_a, for the reaction by making use of the Arrhenius equation:

$$\log_{10} k = \frac{-E_a}{2.30RT} + \text{constant} \tag{4}$$

In this equation, k is the rate constant at the Kelvin temperature T, E_a is the activation energy, and R is the gas constant. By plotting $\log_{10} k$ against $1/T$ we should obtain, by Equation 4, a straight line whose slope equals $-E_a/2.30R$. From the slope of that line we can easily calculate the activation energy.

EXPERIMENTAL PROCEDURE

A. Dependence of Reaction Rate on Concentration. In Table 16.1 we have summarized the reagent volumes to be used in carrying out the several reactions whose rates we need to know in order to find the general rate law for Reaction 1. First, measure out 100 ml of each of the listed reagents (except H_2O) into clean, labeled flasks or beakers. Use these reagents in your reaction mixtures.

TABLE 16.1   REACTION MIXTURES AT ROOM TEMPERATURE
(REAGENT VOLUMES IN ML)

Reaction Mixture	Reaction Flask I (250 ml)			Reaction Flask II (125 ml)	
	0.010 M KI	0.001 M $Na_2S_2O_3$	H_2O	0.040 M $KBrO_3$	0.10 M HCl
1	10	10	10	10	10
2	20	10	0	10	10
3	10	10	0	20	10
4	10	10	0	10	20
5	8	10	12	5	15

Note: For all mixtures, add 3 or 4 drops of starch indicator to Reaction Flask II.

The actual procedure for each reaction mixture will be much the same, and we will describe it now for Reaction Mixture 1.

Since there are several reagents to mix, and since we don't want the reaction to start until we are ready, we will put some of the reagents into one flask and the rest into another, selecting them so that no reaction occurs until the contents of the two flasks are mixed. Using a 10-ml graduated cylinder to measure volumes, measure out 10 ml 0.010 M KI, 10 ml 0.001 M $Na_2S_2O_3$, and 10 ml distilled water into a 250-ml Erlenmeyer flask (Reaction Flask I). Then measure out 10 ml 0.040 M $KBrO_3$ and 10 ml 0.10 M HCl into a 125-ml Erlenmeyer flask (Reaction Flask II). To Flask II add 3 or 4 drops of starch indicator solution.

Pour the contents of Reaction Flask II into Reaction Flask I and swirl the solutions to mix them thoroughly. Note the time at which the solutions were mixed. Continue swirling the solution. It should turn blue in less than 2 minutes. Note the time at the instant that the blue color appears. Record the temperature of the blue solution to 0.2°C.

Repeat the procedure with the other mixtures in Table 16.1. *Don't forget to add the indicator* before mixing the solutions in the two flasks. The reaction flasks should be rinsed with distilled water between runs. When measuring out reagents, rinse the graduated cylinder with distilled water after you have added the reagents to Reaction Flask I, and before you measure out the reagents for Reaction Flask II. Try to keep the temperature just about the same in all the runs. Repeat any experiments that did not appear to proceed properly.

B.   Dependence of Reaction Rate on Temperature.   In this part of the experiment, the reaction will be carried out at several different temperatures, using Reaction Mixture 1 in all cases. The temperatures we will use will be about 20°C, 40°C, 10°C, and 0°C.

We will take the time at about 20°C to be that for Reaction Mixture 1 as determined at room temperature. To determine the time at 40°C proceed as follows. Make up Reaction Mixture 1 as you did in Part A, including the indicator. However, instead of mixing the solutions in the two flasks at room temperature, put the flasks into water at 40°C, drawn from the hot water tap into one or more large beakers. Check to see that the water is indeed at about 40°C, and leave the flasks in the water for several minutes to bring them to the proper temperature. Then mix the two solutions, noting the time of mixing. Continue swirling the reaction flask in the warm water. When the color change occurs note the time and the temperature of the solution in the flask.

Repeat the experiment at about 10°C, cooling all the reactants in water at that temperature before starting the reaction. Record the time required for the color to change and the final temperature of the reaction mixture. Repeat once again at about 0°C, this time using an ice-water bath to cool the reactants.

C.   Dependence of the Reaction Rate on the Presence of Catalyst (Optional).   Some ions have a pronounced catalytic effect on the rates of many reactions in water solution. Observe the effect on this reaction by once again making up Reaction Mixture 1. Before mixing, add 1 drop 0.5 M $(NH_4)_2MoO_4$, ammonium molybdate, and a few drops of starch indicator to Reaction Flask II. Swirl the flask to mix the catalyst thoroughly. Then mix the solutions, noting the time required for the color to change.

DATA AND CALCULATIONS: Rates of Chemical Reactions, II.
A Clock Reaction

A. Orders of the Reaction. Rate Constant Determination

Reaction: $6\,I^-(aq) + BrO_3^-(aq) + 6\,H^+(aq) \rightarrow 3\,I_2(aq) + Br^-(aq) + 3\,H_2O$ (1)

$$rate = k\,(I^-)^m(BrO_3^-)^n(H^+)^p = -\frac{\Delta(BrO_3^-)}{t} \qquad (2)$$

In all the reaction mixtures used in the experiment, the color change occurred when a constant predetermined number of moles of BrO_3^- had been used up by the reaction. The color "clock" allows you to measure the *time required* for this *fixed number of moles of* BrO_3^- *to react*. The rate of each reaction is determined by the time t required for the color to change; since in Equation 2 the change in concentration of BrO_3^- ion, $\Delta(BrO_3^-)$, is the same in each mixture, the relative rate of each reaction is inversely proportional to the time t. Since we are mainly concerned with relative rather than absolute rate, we will for convenience take all relative rates as being equal to $1000/t$. Fill in the data table below, first calculating the relative reaction rate for each mixture.

Reaction mixture	Time t(sec) for color to change	Relative rate of reaction, $1000/t$	(I^-)	Reactant concentrations in reacting mixture (M) (BrO_3^-)	(H^+)	Temp. in °C
1	_____	_____	0.0020	_____	_____	_____
2	_____	_____	_____	_____	_____	_____
3	_____	_____	_____	_____	_____	_____
4	_____	_____	_____	_____	_____	_____
5	_____	_____	_____	_____	_____	_____

The reactant concentrations in the reaction mixture are *not* those of the stock solutions, since the reagents were diluted by the other solutions. The final volume of the reaction mixture is 50 ml in all cases. Since the number of moles of reactant does not change on dilution we can say, for example, for I^- ion, that

$$no.\ moles\ I^- = (I^-)_{stock} \times V_{stock} = (I^-)_{mixture} \times V_{mixture}$$

For Reaction Mixture 1, $(I^-)_{stock} = 0.010\ M,\ V_{stock} = 10\ ml,\ V_{mixture} = 50\ ml$

Therefore, $(I^-)_{mixture} = \dfrac{0.010\ M \times 10\ ml}{50\ ml} = 0.0020\ M$

Calculate the rest of the concentrations in the data table by the same approach.

Continued on following page   **131**

Determination of the Orders of the Reaction

Given the data in the table, the problem is to find the order for each reactant and the rate constant for the reaction. Since we are dealing with relative rates, we can modify Equation 2 to read as follows:

$$\text{relative rate} = k'(\text{I}^-)^m(\text{BrO}_3^-)^n(\text{H}^+)^p \tag{5}$$

We need to determine the relative rate constant k' and the orders m, n, and p in such a way as to be consistent with the data in the table.

The solution to this problem is quite simple, once you make a few observations on the reaction mixtures. Each mixture (2 to 4) differs from Reaction Mixture 1 in the concentration of one species (see data table). This means that for any pair of mixtures that includes Reaction Mixture 1, there is only one concentration that changes. From the ratio of the relative rates for such a pair of mixtures we can find the order for the reactant whose concentration was changed. Proceed as follows:

Write Equation 5 below for Reaction Mixtures 1 and 2, substituting the relative rates and the concentrations of I^-, BrO_3^-, and H^+ ions from the data table you have just completed.

Relative rate 1 = _____ = $k'($ $)^m ($ $)^n ($ $)^p$

Relative rate 2 = _____ = $k'($ $)^m ($ $)^n ($ $)^p$

Divide the first equation by the second, noting that nearly all the terms cancel out! The result is simply

$$\frac{\text{Relative rate 1}}{\text{Relative rate 2}} =$$

If you have done this properly, you will have an equation involving only m as an unknown. Solve this equation for m, the order of the reaction with respect to I^- ion.

$$m = \text{_____} \text{ (nearest integer)}$$

Applying the same approach to Reaction mixtures 1 and 3, find the value of n, the order of the reaction with respect to BrO_3^- ion.

Relative rate 1 = _____ = $k'($ $)^m ($ $)^n ($ $)^p$

Relative rate 3 = _____ = $k'($ $)^m ($ $)^n ($ $)^p$

Dividing one equation by the other:

$$=$$

$$n = \text{_____}$$

Continued on following page

Now that you have the idea, apply the method once again, this time to Reaction Mixtures 1 and 4, and find p, the order with respect to H^+ ion.

Relative rate 4 = $k'($ $)^m ($ $)^n ($ $)^p$

Dividing the equation for Relative rate 1 by that for Relative rate 4, we get

$$=$$

$$p = \underline{\hspace{2cm}}$$

Having found m, n, and p (nearest integers), the relative rate constant, k', can be calculated by substitution of m, n, p, and the known rates and reactant concentrations into Equation 5. Evaluate k' for Reaction Mixtures 1 to 4.

Reaction   1   2   3   4

k' _____ _____ _____ _____ k'_{ave} _____

Why should k' have nearly the same value for each of the above Reactions?

Using k'_{ave} in Equation 5, predict the relative rate and time t for Reaction Mixture 5. Use the concentrations in the table.

relative rate$_{pred}$ _____ t_{pred} _____ t_{obs} _____

B.   The Effect of Temperature on Reaction Rate: The Activation Energy. To find the activation energy for the reaction it will be helpful to complete the table on the next page. The dependence of the rate constant, k', for a reaction is given by Equation 4:

$$\log_{10} k' = \frac{-E_a}{2.3RT} + \text{constant} \tag{4}$$

Since the reactions at the different temperatures all involve the same reactant concentrations, the rate constants, k', for two different mixtures will have the same ratio as the reaction rates themselves for the two mixtures. This means that in the calculation of E_a, we can use the observed relative rates instead of rate constants. Proceeding as before, calculate the relative rates of reaction in each of the mixtures and enter these values in (c) below. Take the $\log_{10}$ rate for each mixture and enter these values in (d). To set up the terms in $1/T$, fill in (b), (e), and (f) in the table.

Continued on following page

Approximate temperature in °C

	20	40	10	0
(a) Time t in seconds for color to appear	———	———	———	———
(b) Temperature of the reaction mixture in °C	———	———	———	———
(c) Relative rate = 1000/t	———	———	———	———
(d) Log$_{10}$ of relative rate	———	———	———	———
(e) Temperature T in K	———	———	———	———
(f) 1/T, K^{-1}	———	———	———	———

To evaluate E_a, make a graph of log relative rate vs. $1/T$ on the graph paper provided. Find the slope of the line obtained by drawing the best straight line through the experimental points. Note that each square on the abscissa equals 0.00005 and each square on the ordinate equals 0.2.

Slope = —————

The slope of the line equals $-E_a/2.3\,R$, where $R = 8.31$ joules/mole K if E_a is to be in joules per mole. Find the activation energy, E_a, for the reaction.

E_a = ————— joules

C. (Optional) Effect of a Catalyst on Reaction Rate

	Reaction 1	Catalyzed Reaction 1
Time for color to appear (seconds)	—————	—————

Would you expect the activation energy, E_a, for the catalyzed reaction to be greater than, less than, or equal to the activation energy for the uncatalyzed reation? Why?

RATES OF CHEMICAL REACTIONS, II. A CLOCK REACTION
(Data and Calculations)

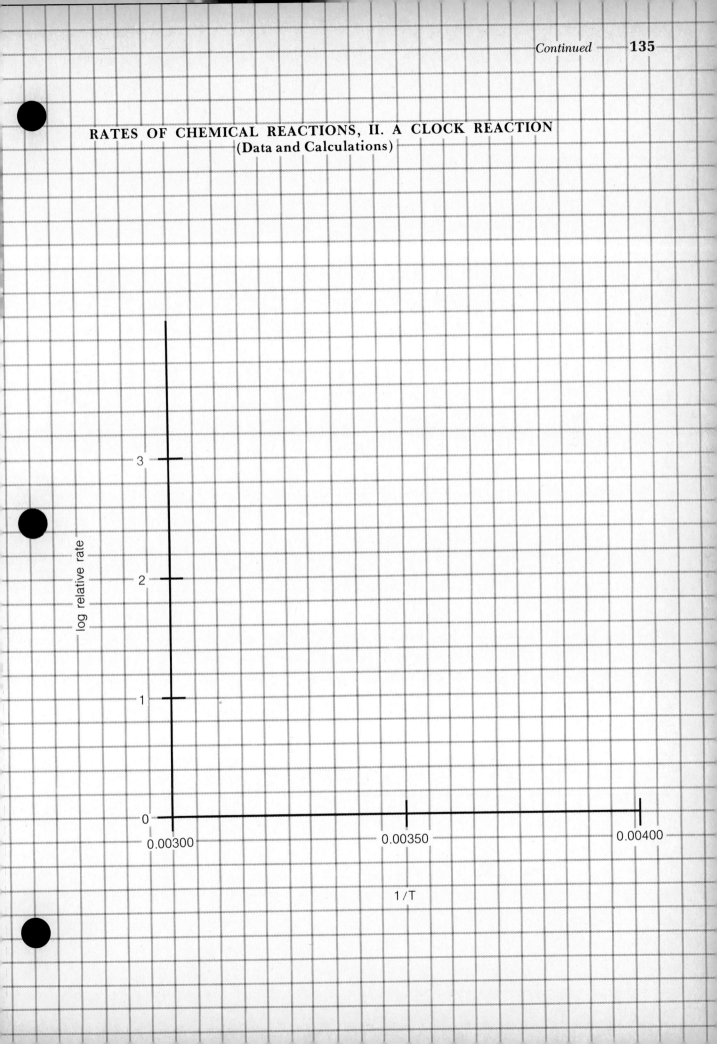

ADVANCE STUDY ASSIGNMENT: Rate of Reactions, II.
                                    A Clock Reaction

1.   A student studied the clock reaction described in this experiment. She set up Reaction Mixture 3 by mixing 10 ml 0.010 M KI, 10 ml 0.001 M $Na_2S_2O_3$, 20 ml 0.040 M $KBrO_3$, and 10 ml 0.10 M HCl using the procedure given. It took about 40 seconds for the color to turn blue.

     a. She found the concentrations of each reactant in the reacting mixture by realizing that the number of moles of each reactant did not change when that reactant was mixed with the others, but that its concentration did. For any reactant A,

$$\text{no. moles A} = M_{A\ stock} \times V_{stock} = M_{A\ mixture} \times V_{mixture}$$

     The volume of the mixture was 50 ml. Revising the above equation, she obtained

$$M_{A\ mixture} = M_{A\ stock} \times \frac{V_{stock}(ml)}{50\ ml}$$

     Find the concentrations of each reactant by using the above equation.

              $(I^-) =$ _____ M; $(BrO_3^-) =$ _____ M; $(H^+) =$ _____ M

     b. What was the relative rate of the reaction $(1000/t)$? _____

     c. Knowing the relative rate of reaction for Mixture 3 and the concentrations of I^-, BrO_3^-, and H^+ in that mixture, she was able to set up Equation 5 for the relative rate of the reaction. The only quantities that remained unknown were k', m, n, and p. Set up Equation 5 as she did, presuming she did it properly.

2.   For Reaction Mixture 1 the student found that 75 seconds were required. On dividing Equation 5 for Reaction Mixture 1 by Equation 5 for Reaction Mixture 3, and after canceling out the common terms (k', terms in (I^-) and (H^+)), she got the following equation:

$$\frac{13.3}{25} = \left(\frac{0.004}{0.008}\right)^n = \left(\frac{1}{2}\right)^n$$

Recognizing that 13.3/25 is about equal to $\frac{1}{2}$, she obtained an approximate value for n. What was that value?

                                                  $n =$ _____

Continued on following page   **137**

By taking logarithms of both sides of the equation, she got an exact value for n. What was that value?

$$n = \underline{\hspace{2cm}}$$

Since orders of reactions are often integers, she reported her approximate value as the order of the reaction with respect to BrO_3^-.

EXPERIMENT

17 • Some Nonmetals and their Compounds—Preparation and Properties

Some of the most commonly encountered chemical substances are nonmetallic elements or their simple compounds. O_2 and N_2 in the air, CO_2 produced by combustion of oil, coal, or wood, and H_2O in the rivers, lakes, and air are typical of such substances. Substances containing nonmetallic atoms, whether elementary or compound, are all molecular, reflecting the covalent bonding that holds their atoms together. They are often gases, due to weak intermolecular forces. However, with high molecular masses, hydrogen bonding, or macromolecular structures one finds liquids, like H_2O and Br_2, and solids, like I_2 and graphite.

Several of the common nonmetallic elements and some of their gaseous compounds can be prepared by simple reactions. In this experiment you will prepare some typical examples of such substances and examine a few of their characteristic properties.

EXPERIMENTAL PROCEDURE
WEAR YOUR SAFETY GLASSES WHILE PERFORMING THIS EXPERIMENT

In several of the experiments you will be doing you will prepare gases. In general we will not describe in detail what you should observe, so perform each preparation carefully and report what you actually observe, not what you think we expect you to observe.

There are several tests we will make on the gases you prepare, and the way each of these should be carried out is summarized below.

Test for Odor. To determine the odor of a gas, first pass your hand across the end of the tube, bringing the gas toward your nose. If you don't detect an odor, sniff near the end of the tube, first at some distance and then gradually closer. Don't just put the tube at the end of your nose immediately and take a deep breath. Some of the gases you will make have no odor, and some will have very impressive ones. Some of the gases are very toxic and, even though we will be making only small amounts, caution in testing for odor is important.

Test for Support of Combustion. A few gases will support combustion, but most will not. To make the test, ignite a wood splint with a Bunsen flame, blow out the flame, and put the glowing, but not burning, splint into the gas in the test tube. If the gas supports combustion, the splint will glow more brightly, or may make a small popping noise. If the gas does not support combustion, the splint will go out almost instantly. You can use the same splint for all the support of combustion tests.

Test for Acid-Base Properties. Many gases are acids. This means that if the gas is dissolved in water it will produce some H^+ ions. A few gases are bases; in water solution such gases produce OH^- ions. Other gases do not interact with water, and are neutral. It is easy to establish the acidic or basic nature of a gas by using a chemical indicator. One of the most common acid-base indicators is litmus, which is red in acidic solution and blue in basic solution. To test whether a gas is an acid, moisten a piece of blue litmus paper with water from your wash bottle, and put the paper down into the test tube in which the gas is present. If the gas is an acid, the paper will turn red (blue to red, "as-sed"). Similarly, to test if the gas is a base, moisten a piece of red litmus paper, and hold it down in the tube. A color change to blue will occur if the gas forms a basic solution (red to blue, "alkaloo"). Since you may have used an acid or a base in making the gas, do not touch the walls of the test tube with the paper. The color change will occur fairly quickly and smoothly over the surface of the paper if the gas is acidic or basic. It is not necessary to use a new piece of litmus for each test. Start with a piece of blue and a piece of red litmus. If you need to regenerate the blue paper, hold it over an open bottle of 6 M NH_3, whose vapor is basic. If you need to make red litmus, hold the moist paper over an open bottle of 6 M acetic acid.

A. Preparation and Properties of Nonmetallic Elements: O_2, N_2, Br_2, I_2

1. OXYGEN, O_2. Oxygen can be easily prepared in the laboratory by the decomposition of H_2O_2, hydrogen peroxide, in aqueous solution. Hydrogen peroxide is not very stable and will break down to water and oxygen gas on addition of a suitable catalyst, particularly MnO_2:

$$2 \, H_2O_2(aq) \xrightarrow{\text{MnO}_2} O_2(g) + 2 \, H_2O \tag{1}$$

Add 1 ml 3% H_2O_2 solution in water to a small test tube. Pick up a very small amount ($\sim$0.1 g) of MnO_2 on the tip of your spatula and add it to the liquid in the tube. Hold your finger over the end of the tube to help confine the O_2. Test the evolved gas for odor. Test the gas for any acid-base properties, using moist blue and red litmus paper. Test the gas for support of combustion. Record your observations.

2. NITROGEN, N_2. Hydrazine, N_2H_4, in water solution will react with CrO_4^{2-} ion to form nitrogen gas and greenish $Cr(OH)_3$:

$$3 \, N_2H_4(aq) + 4 \, CrO_4^{2-}(aq) + 4 \, H_2O \rightarrow 3 \, N_2(g) + 4 \, Cr(OH)_3(s) + 8 \, OH^-(aq) \tag{2}$$

Hydrazine, like H_2O_2, can be decomposed with an MnO_2 catalyst, but the reaction is much slower.

To a small test tube add about 2 ml 0.25 M N_2H_4. Add 6 drops 1 M K_2CrO_4. Confine the gas for about 2 seconds with your finger. Cautiously test the evolved gas for odor. Carry out the tests for acid-base properties and for support of combustion. Report your observations.

3. IODINE, I_2. The halogen elements are most readily prepared from their sodium or potassium salts. The reaction involves an oxidizing agent, which can remove electrons from the halide ions, freeing the halogen. The reaction which occurs when a solution of potassium iodide, KI, is treated with 6 M HCl and a little MnO_2 is

$$2 \, I^-(aq) + 4 \, H^+(aq) + MnO_2(s) \rightarrow I_2(aq) + Mn^{2+}(aq) + 2 \, H_2O \tag{3}$$

At 25°C I_2 is a solid, with relatively low solubility in water, and appreciable volatility. I_2 can be extracted from aqueous solution into organic solvents, particularly trichloroethane, CCl_3CH_3 (TCE). The solid, vapor, and solutions in water and TCE all have characteristic colors.

Put 2 drops 1 M KI into a *regular* test tube. Add 6 drops 6M HCl and a tiny amount of manganese dioxide, MnO_2. Swirl the mixture and note any changes which occur. Put the test tube into a hot water bath made from a 250-ml beaker about half full of water. After a minute or two a noticeable amount of I_2 vapor should be visible above the liquid. Remove the test tube from the water bath and add 10 ml distilled water. Stopper the tube and shake. Note the color of I_2 in the solution. Sniff the vapor above the solution. Decant the liquid into another regular test tube and add 3 ml trichloroethane, CCl_3CH_3. Stopper and shake the tube. Observe the color of the TCE layer and the relative solubility of I_2 in water and TCE. Record your observations.

4. BROMINE, Br_2. Bromine can be made by the same reaction as is used to make iodine, substituting bromide ion for iodide ion. Bromide ion is less easily oxidized than iodide ion. Bromine at 25°C is a liquid. Br_2 can be extracted from water solution into trichloroethane. Its liquid, vapor, and solutions in water and TCE are colored.

To 3 drops 1 M NaBr in a regular test tube add 6 drops 6 M HCl and a small amount of MnO_2

(about the size of a small pea). Swirl to mix the reagents, and observe any changes. Heat the tube in the water bath for a minute or two. Try to detect Br_2 vapor above the liquid by observing its color. Add 10 ml water to the tube, stopper, and shake. Note the color of Br_2 in the solution. Sniff the vapor, cautiously. Decant the liquid into a regular test tube. Add 3 ml TCE, stopper and shake. Note the color of Br_2 in TCE. Record your observations.

B. Preparation and Properties of Some Nonmetallic Oxides: CO_2, SO_2, NO, and NO_2

1. CARBON DIOXIDE, CO_2. Carbon dioxide, like several nonmetallic oxides, is easily made by treating an oxyanion with an acid. With carbon dioxide the oxyanion is CO_3^{2-}, carbonate ion, which is present in solutions of carbonate salts, such as Na_2CO_3. The reaction is

$$CO_3^{2-}(aq) + 2\ H^+(aq) \rightarrow H_2CO_3(aq) \rightarrow CO_2(g) + H_2O \tag{4}$$

Carbon dioxide is not very soluble in water, and on acidification, carbonate solutions will tend to effervesce as CO_2 is liberated. Nonmetallic oxides in solution are often acidic but never basic.

To 1 ml 1 M Na_2CO_3 in a small test tube add 6 drops 3 M H_2SO_4. Test the gas for odor, acidic properties, and ability to support combustion.

2. SULFUR DIOXIDE, SO_2. Sulfur dioxide is readily prepared by acidification of a solution of sodium sulfite, containing sulfite ion, SO_3^{2-}. As you can see, the reaction which occurs is very similar to that with carbonate ion.

$$SO_3^{2-}(aq) + 2\ H^+(aq) \rightarrow H_2SO_3(aq) \rightarrow SO_2(g) + H_2O \tag{5}$$

Sulfur dioxide is considerably more soluble in water than is carbon dioxide. Some effervescence may be observed on acidification of concentrated sulfite solutions, which increases if the solution is heated in a water bath.

To 1 ml 1 M Na_2SO_3 in a small test tube add 6 drops 3 M H_2SO_4. *Cautiously* test the evolved gas for odor. Test its acidic properties and its ability to support combustion. Put the test tube into the water bath for a few seconds to see if effervescence occurs if the solution is hot.

3. NITROGEN DIOXIDE, NO_2, AND NITRIC OXIDE, NO. If a solution containing nitrite ion, NO_2^-, is treated with acid, two oxides are produced, NO_2 and NO. In solution these gases are combined in the form of N_2O_3, which is colored. When these gases come out of solution, the mixture contains NO and NO_2; the latter is colored. NO is colorless and reacts readily with oxygen in the air to form NO_2. The preparation reaction is

$$2\ NO_2^-(aq) + 2\ H^+(aq) \rightarrow N_2O_3(aq) + H_2O \rightarrow NO(g) + NO_2(g) + H_2O \tag{6}$$

To 1 ml of 1 M KNO_2 in a small test tube, add 6 drops 3 M H_2SO_4. Swirl the mixture for a few seconds and note the color of the solution. Warm the tube in the water bath for a few seconds to increase the rate of gas evolution. Note the color of the gas that is given off. Cautiously test the odor of the gas. Test its acidic properties and its ability to support combustion. Record your observations.

C. Preparation and Properties of Some Nonmetallic Hydrides: NH_3, H_2S

1. AMMONIA, NH_3. A 6 M solution of NH_3 in water is a common laboratory reagent. You may have been using it in this experiment to make your litmus paper turn blue. Ammonia gas can be made by simply heating 6 M NH_3. It can also be prepared by addition of a strongly basic solution to a solution of an ammonium salt, such as NH_4Cl. The latter solution contains NH_4^+ ion. On treatment with OH^- ion, as in a solution of NaOH, the following reaction occurs:

$$NH_4^+(aq) + OH^-(aq) \rightarrow NH_3(aq) + H_2O \rightarrow NH_3(g) + H_2O \tag{7}$$

The odor of NH_3 is characteristic. NH_3 is very soluble in water, so effervescence is not observed, even on heating concentrated solutions.

Add 1 ml 1 M NH_4Cl to a small test tube. Add 1 ml 6 M NaOH. Swirl the mixture and test the odor of the evolved gas. Put the test tube into the water bath for a few moments to increase the amount of NH_3 in the gas phase. Test the gas with moistened blue and red litmus paper. Test the gas for support of combustion.

2.   HYDROGEN SULFIDE, H_2S.   Hydrogen sulfide can be made by treating some solid sulfides, particularly FeS, with an acid such as HCl or H_2SO_4. With FeS the reaction is

$$FeS(s) + 2\ H^+(aq) \rightarrow H_2S(g) + Fe^{2+}(aq) \tag{8}$$

This reaction was used for many years to make H_2S in the laboratory in courses in qualitative analysis. In this experiment we will employ the method currently used in such courses for H_2S generation. This involves the decomposition of thioacetamide, CH_3CSNH_2, which occurs in solution on treatment with acid and heat. The reaction is

$$CH_3CSNH_2(aq) + 2\ H_2O \rightarrow H_2S(g) + CH_3COO^-(aq) + NH_4^+(aq) \tag{9}$$

The odor of H_2S is notorious. H_2S is moderately soluble in water and, since it is produced reasonably slowly in Reaction 9, there will be little if any effervescence.

To 1 ml of 1 M thioacetamide in a small test tube add 6 drops 3 M H_2SO_4. Put the test tube in a boiling water bath for about 1 minute. There may be some cloudiness due to formation of free sulfur. Carefully smell the gas in the tube. Test the gas for acidic and basic properties and for the ability to support combustion.

D. (Optional).   In this experiment you prepared nine different species containing nonmetallic elements. In each case the source of the species was in solution. In this part of the experiment we will give you an unknown solution which can be used to make one of the nine species. The unknown is one of the solutions you used in this experiment. Identify by suitable tests the species which can be made from your unknown and the substance that is present in your unknown solution.

 Name ______________ Section ______________

DATA AND OBSERVATIONS: Nonmetals and Their Compounds

A.

	Element prepared	Degree of effervescence	Odor	Acid-base tests	Support of combustion test
1.	O_2	______	______	______	______
2.	N_2	______	______	______	______

		Color		
		Vapor	In H_2O	In TCE
3.	I_2	______ ______	______	______
4.	Br_2	______ ______	______	______

B.

	Oxide prepared	Degree of effervescence	Odor	Acid test	Color	Support of combustion test
1.	CO_2	______	______	______	______	______
2.	SO_2	______	______	______	______	______
3.	$NO_2 + NO$	______	______	______	soln ______	______
					gas ______	

C.

	Hydride prepared	Degree of effervescence	Odor	Acid-base tests	Support of combustion test
1.	NH_3	______	______	______	______
2.	H_2S	______	______	______	______

D. (Optional) Properties of unknown:

Nonmetal species which can be made from unknown ______________

Identity of unknown solution ______________

Unknown no. ______________

 Name ______________________ Section ______________

ADVANCE STUDY ASSIGNMENT: Nonmetals and Their Compounds

In this experiment nine species are prepared and studied. For each of the species, list the reagents that are used in its preparation and the reaction that occurs.

	Reagents used	Reaction
1. O_2		
2. N_2		
3. I_2		
4. Br_2		
5. CO_2		
6. SO_2		
7. $NO_2 + NO$		
8. NH_3		
9. H_2S		

18 • Determination of an Unknown Chloride

One of the important applications of precipitation reactions lies in the area of quantitative analysis. Many substances that can be precipitated from solution are so slightly soluble that the precipitation reaction by which they are formed can be considered to proceed to completion. Silver chloride is an example of such a substance.

$$AgCl(s) \rightleftharpoons Ag^+(aq) + Cl^-(aq) \qquad K_{sp} = [Ag^+][Cl^-] = 1.6 \times 10^{-10}$$

Although silver chloride is in chemical equilibrium with its ions in solution, the equilibrium constant K_{sp} for the reaction is so low that if AgCl is precipitated by the addition of a solution containing Ag^+ ion to one containing Cl^- ion, essentially all the Ag^+ added will precipitate as AgCl until essentially all the Cl^- is used up. When the amount of Ag^+ added to the solution is equal to the amount of Cl^- originally present, the precipitation of Cl^- ion will be, for all practical purposes, complete.

A convenient method for chloride analysis using AgCl has been devised. A solution of $AgNO_3$ is added to a chloride solution just to the point where the number of moles of Ag^+ added is equal to the number of moles of Cl^- initially present. We analyze for Cl^- by simply measuring how many moles of $AgNO_3$ are required. Surprisingly enough, this measurement is rather easily made by an experimental procedure called a *titration*.

In the titration a solution of $AgNO_3$ of known concentration (in moles $AgNO_3$ per liter of solution) is added from a calibrated buret to a solution containing a measured amount of unknown. The titration is stopped when a color change occurs in the solution, indicating that stoichiometrically equivalent amounts of Ag^+ and Cl^- are present. The color change is caused by a chemical reagent, called an indicator, which is added to the solution at the beginning of the titration.

The volume of $AgNO_3$ solution that has been added up to the time of the color change can be measured accurately with the buret, and the number of moles of Ag^+ added can be calculated from the known concentration of the solution.

In the Mohr method for the volumetric analysis of chloride, which we will employ in this experiment, the indicator used is K_2CrO_4. The chromate ion present in solutions of this substance will react with silver ion to form a red precipitate of Ag_2CrO_4. This precipitate will form as soon as $[Ag^+]^2 \times [CrO_4^{2-}]$ exceeds the solubility product of Ag_2CrO_4, which is about 1×10^{-12}. Under the conditions of the titration, the Ag^+ added to the solution reacts preferentially with Cl^- until that ion is essentially quantitatively removed from the system, at which point Ag_2CrO_4 begins to precipitate and the solution color changes from yellow to buff. The end point of the titration is that point at which the color change is first observed.

In this experiment, solutions containing an unknown concentration of chloride will be titrated with a standardized solution of $AgNO_3$, and the volumes of $AgNO_3$ solution required to reach the end point of each titration will be measured. Given the molarity of the $AgNO_3$

$$\text{no. of moles } Ag^+ = \text{no. of moles } AgNO_3 = M_{AgNO_3} \times V_{AgNO_3} \qquad (1)$$

where the volume of $AgNO_3$ is expressed in liters and the molarity M_{AgNO_3} is in moles per liter of solution. At the end point of the titration,

$$\text{no. of moles } Ag^+ \text{ added} = \text{no. of moles } Cl^- \text{ present in unknown} \qquad (2)$$

The concentration of chloride ion in moles per liter is simply equal to the number of moles of Cl^- ion present in the titrated sample divided by the volume of that sample in liters:

$$[Cl^-] = \frac{\text{no. of moles } Cl^- \text{ present in titrated sample}}{\text{volume of titrated sample in liters}} \tag{3}$$

In our titrations the volume of the titrated sample will be the same in all cases and equal to 10.0 ml, or 0.0100 liter.

If we wish to find the number of grams of chloride ion in the titrated sample we multiply the number of moles by the mass of one mole of chloride ion, which is numerically equal to the atomic mass of chlorine, 35.45.

$$\text{no. of grams } Cl^- = \text{no. of moles } Cl^- \times \frac{35.45 \text{ g}}{1 \text{ mole}} \tag{4}$$

EXPERIMENTAL PROCEDURE

Take a clean, *dry* 250-ml Erlenmeyer flask to the stockroom. You will be given about 50 ml of an unknown chloride solution, a buret, and a 10-ml pipet.

Go to the reagent shelf and, using the graduated cylinder there, measure out about 100 ml of standardized $AgNO_3$ solution into another clean, *dry* 250-ml Erlenmeyer flask. This will be your total supply for the entire experiment, so do not waste it. Clean your buret thoroughly with soap solution and rinse it with distilled water. Pour three successive 2- or 3-ml portions of the $AgNO_3$ solution into the buret and tip it back and forth to rinse the inside walls. Allow the $AgNO_3$ to drain out of the buret tip completely each time. Fill the buret with the $AgNO_3$ solution. Open the stopcock momentarily to flush any air bubbles out of the tip of the buret. Be sure that your stopcock fits snugly and that the buret does not leak.

Pipet 10 ml of your unknown chloride solution into a clean, but not necessarily dry, 125-ml Erlenmeyer flask. This is the sample you will be titrating. To that sample add about 25 ml distilled water, to give you enough liquid volume to work with, and 3 drops of 1 M K_2CrO_4, which will give the solution a yellow color.

Read the initial buret level to 0.02 ml. You may find it useful when making readings to put a white card marked with a thick, black stripe behind the meniscus. If the black line is held just below the level to be read, its reflection in the surface of the meniscus will help you obtain a very accurate reading. Begin to add the $AgNO_3$ solution to the chloride solution in the Erlenmeyer flask. A white precipitate of AgCl will form immediately, and the amount will increase during the course of the titration. At the beginning of the titration, you can add the $AgNO_3$ fairly rapidly, a few ml at a time, swirling the flask as best you can to mix the solution. You will find that at the point where the $AgNO_3$ hits the solution, there will be a red trace of Ag_2CrO_4, which disappears when you stop adding nitrate and swirl the flask. As you proceed with the titration, the red color will persist more and more, since the amount of excess chloride ion, which reacts with the Ag_2CrO_4 to form AgCl, will slowly decrease. Gradually decrease the rate at which you add $AgNO_3$ as the red color becomes stronger. At some stage you may find it convenient to set your buret stopcock to deliver $AgNO_3$ slowly, drop by drop, while you swirl the flask. When you are near the end point, add the $AgNO_3$ drop by drop, swirling between drops. The end point of the titration is that point where the mixture first takes on a permanent reddish-yellow or buff color which does not revert to pure yellow on swirling. If you are careful, you can hit the end point within a few drops of $AgNO_3$. When you have reached the end point, stop the titration and record the buret level.

Pour the solution you have just titrated into another 125-ml Erlenmeyer flask or into a 400-ml beaker. To that solution add a few ml of your unknown chloride solution. The color of the mixture should revert to the original yellow. Use the color of this mixture as a reference against which you can compare your samples in the remaining titrations.

Using distilled water, rinse out the 125-ml Erlenmeyer flask in which you carried out the titration. Then pipet another 10-ml sample of unknown (sometimes called an aliquot) into the flask. Add about 25 ml distilled water and the K_2CrO_4 indicator. Refill your buret, take a volume reading, and titrate the solution as before. This titration should be more accurate than the first, for two reasons. The sample volume to be titrated is the same as before, so the volume of $AgNO_3$ will be nearly the same as before. In addition, you have a reference for color comparison which should make it easier to recognize when a color change has occurred.

Titrate the third sample as you did the second. With care it should be possible to obtain volumes of $AgNO_3$ which agree to within less than 1% in the last two titrations.

Silver nitrate is very expensive. Pour all titrated solutions and any $AgNO_3$ solution remaining in your buret or flask into the waste bottles provided.

Name _____ Section _____

DATA AND CALCULATIONS: Determination of an Unknown Chloride

Unknown sample no. _____

Molarity of standardized $AgNO_3$ solution _____ M

	I	II	III
Initial buret reading	_____ ml	_____ ml	_____ ml
Final buret reading	_____ ml	_____ ml	_____ ml
Volume of $AgNO_3$ used in titration of 10-ml sample of unknown	_____ ml	_____ ml	_____ ml
No. of moles $AgNO_3$ used	_____ moles	_____ moles	_____ moles
No. of moles Cl^- present in 10-ml sample	_____ moles	_____ moles	_____ moles
Molarity of Cl^- in unknown	_____ M	_____ M	_____ M

Value of molarity that you   _____ M
wish to report

If you didn't select the average molarity as your reported value, state your reasons:

No. of grams of Cl^- present in   _____ g
10-ml sample of unknown

No. of grams of Cl^- present in   _____ g
1 liter of unknown

ADVANCE STUDY ASSIGNMENT: Determination of an Unknown Chloride

1. A student pipets 10 ml of an unknown chloride solution into a 125-ml Erlenmeyer flask. He adds 25 ml water and 3 drops 1 M K_2CrO_4. He cleans his buret and fills it with 0.02268 M $AgNO_3$. The initial volume reading on the buret is 1.23 ml. He titrates the unknown solution until the color just turns from yellow to buff and finds that the volume reading on the buret is 32.16 ml.

    a. What volume of $AgNO_3$ did he use?

                                        _____ ml = _____ liters

    b. How many moles of $AgNO_3$ were in that volume (Eq. 1)?

                                          _____ moles

    c. How many moles of Ag^+ were in that volume of solution?   _____ moles

    d. How many moles of Cl^- were there in the sample that was titrated (Eq. 2)?

                                          _____ moles

    e. How many grams of Cl^- were in the titrated sample (Eq. 4)?

                                          _____ grams

    f. What volume of unknown solution was used in the titration?

                                          _____ ml = _____ liters

    g. What was the concentration of Cl^- in the *unknown* in moles per liter (Eq. 3).

                                          _____ M

EXPERIMENT

19 • Molecular Mass Determination by Depression of the Freezing Point

In an earlier experiment you observed the change of vapor pressure of a liquid as a function of temperature. If a nonvolatile solid compound (the solute) is dissolved in a liquid, the vapor pressure of the liquid solvent is lowered. This decrease in the vapor pressure of the solvent results in other easily observable physical changes; the boiling point of the solution is higher than that of the pure solvent and the freezing point is lower.

Many years ago chemists observed that at low solute concentrations the changes in the boiling point, the freezing point, and the vapor pressure of a solution are all proportional to the amount of solute that is dissolved in the solvent. These three properties are collectively known as colligative properties of solutions. The colligative properties of a solution depend only on the number of solute particles present in a given amount of solvent and not on the kind of particles dissolved.

When working with boiling point elevations or freezing point depressions of solutions, it is convenient to express the solute concentration in terms of its molality m defined by the relation:

$$\text{molality of } A = m_A = \frac{\text{no. of moles } A \text{ dissolved}}{\text{no. of kg solvent in the solution}}$$

For this unit of concentration, the boiling point elevation, $T_b - T_b^\circ$ or ΔT_b, and the freezing point depression, $T_f^\circ - T_f$ or ΔT_f, in °C at low concentrations are given by the equations:

$$\Delta T_b = k_b m \qquad \Delta T_f = k_f m \tag{1}$$

where k_b and k_f are characteristic of the solvent used. For water, $k_b = 0.52$ and $k_f = 1.86$. For benzene, $k_b = 2.53$ and $k_f = 5.10$.

One of the main uses of the colligative properties of solutions is in connection with the determination of the molecular masses of unknown substances. If we dissolve a known amount of solute in a given amount of solvent and measure ΔT_b or ΔT_f of the solution produced, and if we know the appropriate k for the solvent, we can find the molality and hence the molar mass, GMM, of the solute. In the case of the freezing point depression, the relation would be:

$$\Delta T_f = k_f m = k_f \times \frac{\text{no. moles solute}}{\text{no. kg solvent}} = k_f \times \frac{\left(\dfrac{\text{no. g solute}}{\text{GMM solute}}\right)}{\text{no. kg solvent}} \tag{2}$$

In this experiment you will be asked to estimate the molar mass of an unknown solute, using this equation. The solvent used will be paradichlorobenzene, which has a convenient melting point and a relatively large value for k_f, 7.10. The freezing points will be obtained by studying the rate at which liquid paradichlorobenzene and some of its solutions containing the unknown cool in air.

When a pure substance which melts at, say, 60°C is heated to 70°C, where it will be completely liquid, and then allowed to cool in air, the temperature of the sample will vary with time, as in Figure 19.1. Initially the temperature will fall quite rapidly. When the freezing point is reached, solid will begin to form, and the temperature will tend to hold steady until the sample is all solid. The freezing point of the pure liquid is the constant temperature observed while the liquid is freezing to a solid.

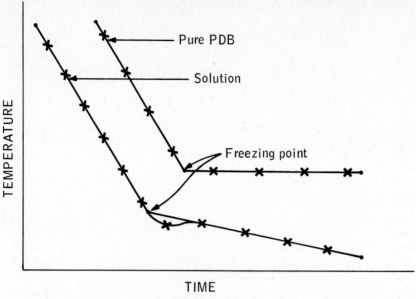

Figure 19.1

The cooling behavior of a solution is somewhat different from that of a pure liquid, and is also shown in Figure 19.1. The temperature at which the solution begins to freeze is lower than for the pure solvent. In addition, there is a slow gradual fall in temperature as freezing proceeds. The best value for the freezing point of the solution is obtained by drawing two straight lines connecting the points on the temperature-time graph. The first line connects points where the solution is all liquid. The second line connects points where solid and liquid coexist. The point where the two lines intersect is the freezing point of the solution. With both the pure liquid and solutions, at the time when solid first appears, the temperature may fall below the freezing point and then come back up to it as solid forms. The effect is called supercooling, and is shown in Figure 19.1. When drawing the straight line in the solid-liquid region of the graph, ignore points where supercooling was observed. In order to establish the proper straight line in the solid-liquid region it is necessary to record the temperature until the trend with time is smooth and clearly established.

EXPERIMENTAL PROCEDURE

A. Determination of the Freezing Point of Paradichlorobenzene. From the stockroom obtain a stopper fitted with a sensitive thermometer and a glass stirrer, a large test tube, and a sample of solid unknown. **Remember that the thermometer is both fragile and expensive, so handle it with due care.** Weigh the test tube on a top loading or triple beam balance to 0.01 g. Add about 30 g of paradichlorobenzene, PDB, to the test tube and weigh again to the same precision.

Fill your 600 ml beaker almost full of hot water from the faucet. Support the beaker on an iron ring and wire gauze on a ring stand and heat the water with a Bunsen burner. Clamp the test tube to the ring stand and immerse the tube in the water as far as is convenient (see Fig. 19.2). *CAUTION:* The vapor of PDB is toxic, like that of most organic compounds. Avoid breathing the vapor.

When the water gets above the melting point of the PDB, it will begin to melt. After most of the solid appears to have melted, clamp the thermometer-stirrer assembly as in Figure 19.2, adjusting the level of the thermometer so that the bulb is about 1 cm above the bottom of the tube, well down into the melt. Stir to melt the remaining solid PDB. When the PDB is at about 65°C, stop heating. Carefully lower the iron ring and water bath and put the beaker of hot water on the lab bench well away from the test tube. Dry the outside of the test tube with a towel.

Record the temperature of the paradichlorobenzene as it cools in the air. Stir the liquid slowly but continuously to minimize supercooling. *Start readings at about 60° C and note the temperature*

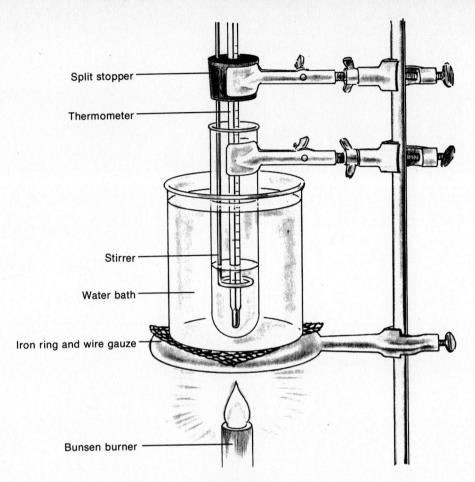

Split stopper

Thermometer

Stirrer

Water bath

Iron ring and wire gauze

Bunsen burner

Figure 19.2

every 30 seconds for 8 minutes or until the liquid has solidified to the point that you are no longer able to stir it. Near the melting point you will begin to observe crystals of PDB in the liquid, and these will increase in amount as cooling proceeds.

B.   Determination of the Molecular Mass of an Unknown Compound.   Weigh your unknown in its container to 0.001 g. Pour about half of the sample (about 2 g) into your test tube of PDB and reweigh the container. Be careful with both weighings.

Heat the test tube in the water bath until the PDB is again melted and the solid unknown is dissolved. Make sure that all the PDB on the walls is melted; stir well to mix the unknown with the PDB thoroughly. When the melt has reached about 65°C, remove the bath, dry the test tube, and let it cool as before. Start readings at about 58°C and continue to take readings, with stirring, for 8 minutes.

The dependence of temperature on time with the solution will be similar to that observed for pure PDB, except that the first crystals will appear at a lower temperature, and the temperature of the solid-solution system will gradually fall as cooling proceeds. There may be some supercooling, as evidenced by a rise in temperature shortly after the first appearance of PDB crystals.

Add the rest of your unknown (about 2 g) to the PDB solution and again weigh the container. Melt the PDB as before, heating it to about 60°C before removing the water bath. Start temperature-time readings at about 55°C. Repeat the entire procedure described above.

　　　　　　　　　　Name _____ Section _____

DATA AND CALCULATIONS: Molecular Mass Determination by Freezing Point Depression

Mass of large test tube　　　　　　　　　　　　　　　　　　　_____g

Mass of test tube plus about 30 g of paradichlorobenzene　　　　_____g

Mass of container plus unknown　　　　　　　　　　　　　　_____g

Mass of container less Sample I　　　　　　　　　　　　　　_____g

Mass of container less Sample II　　　　　　　　　　　　　_____g

Time-temperature Readings:

Temperature, °C

Time (minutes)	Paradichloro-benzene	Solution I	Solution II
0	_____	_____	_____
½	_____	_____	_____
1	_____	_____	_____
1½	_____	_____	_____
2	_____	_____	_____
2½	_____	_____	_____
3	_____	_____	_____
3½	_____	_____	_____
4	_____	_____	_____
4½	_____	_____	_____
5	_____	_____	_____
5½	_____	_____	_____

Continued on following page　　**159**

6	_____	_____	_____
6½	_____	_____	_____
7	_____	_____	_____
7½	_____	_____	_____
8	_____	_____	_____

Estimation of Freezing Point

On the graph paper provided, plot your temperature vs. time readings for pure paradichlorobenzene and for each run of the two solutions. To avoid overlapping graphs, add 4 minutes to all observed times in making the graph of the cooling curve for pure paradichlorobenzene. Add 2 minutes to all times for Solution I. Use times as observed for Solution II. Connect the points on each cooling curve with two straight lines, ignoring points involving supercooling. The freezing point is at the point where the two lines intersect.

Freezing points (from graphs)

_____	_____	_____
Pure PDB	Solution I	Solution II

Calculation of Molecular Mass

	Solution I	Solution II
Mass of unknown used (*total* amount in solution)	_____ g	_____ g
Mass of paradichlorobenzene used	_____ g	_____ g
Freezing point of pure paradichlorobenzene	_____ °C	_____ °C
Freezing point of solution	_____ °C	_____ °C
Freezing point depression (total depression)	_____ °C	_____ °C
Total molal concentration of unknown solution (from Eq. 1; $k_f = 7.10$)	_____	_____
Molar mass, GMM, of unknown (from Eq. 2)	_____ g	_____ g
Average molecular mass	_____	
Unknown no.	_____	

Molecular Mass Determination by Freezing Point Depression

(Data and Calculations)

Name

Section

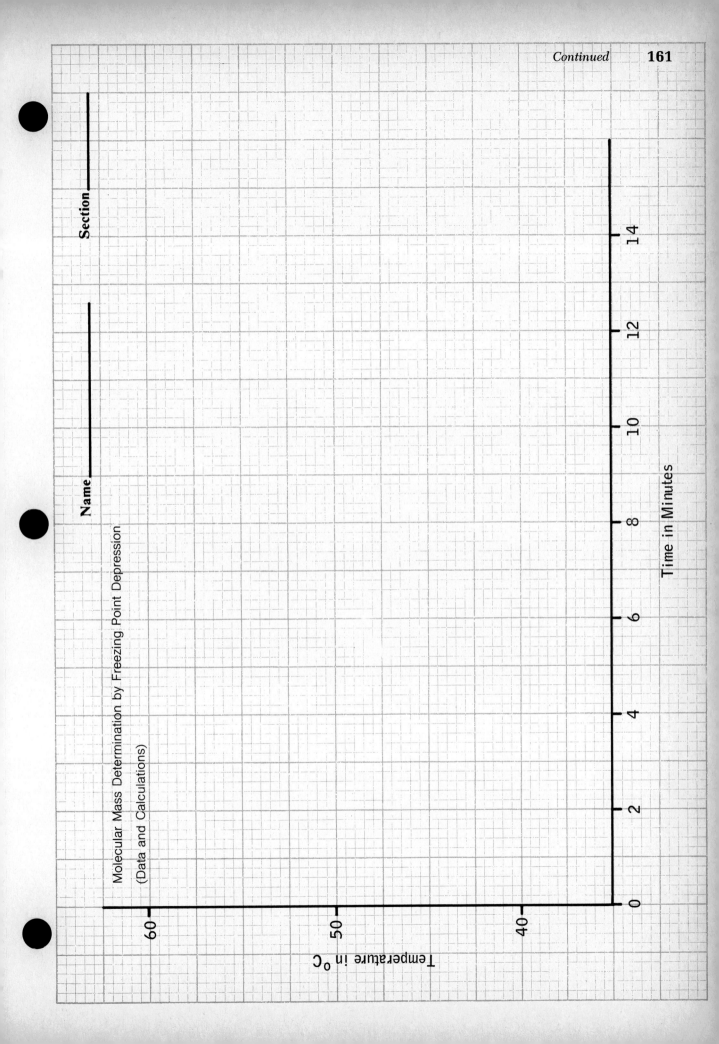

Temperature in $^{\circ}C$

60

50

40

0   2   4   6   8   10   12   14

Time in Minutes

ADVANCE STUDY ASSIGNMENT: Determination of Molecular Mass by
Freezing Point Depression

1. A student determines the freezing point of a solution of 2.00 g of naphthalene in 26.00 g of
paradichlorobenzene. He obtains the following temperature-time readings:

Time (min)	0	$\frac{1}{2}$	1	$1\frac{1}{2}$	2	$2\frac{1}{2}$	3	$3\frac{1}{2}$	4
Temperature (°C)	59.7	58.0	56.5	54.8	53.4	52.1	50.9	49.5	48.4

Time (min)	$4\frac{1}{2}$	5	$5\frac{1}{2}$	6	$6\frac{1}{2}$	7	$7\frac{1}{2}$	8
Temperature (°C)	48.6	48.7	48.6	48.5	48.4	48.3	48.2	48.1

a. Plot these data on the graph paper provided. Note that the first several points lie essentially
on a straight line. Similarly, the last several points lie on another straight line. Draw in
those two lines. The point at which those lines intersect is the freezing point of the solution.
(There is a dip in the observed points at about 4 minutes; this is due to supercooling, and is
to be ignored when drawing the two lines.)

b. What is the freezing point of the solution?   _____ °C

c. What is the freezing point depression, $T_f^\circ - T_f$, or ΔT_f? Take T_f° to   _____ °C
be 53.0°C for paradichlorobenzene.

d. What is the molality, m, of the naphthalene? (Use Eq. 1; $k_f = 7.10$)   _____ m

e. What is the molar mass of naphthalene? Use Equation 2, as modified below:

$$\text{GMM} = \frac{\text{no. g solute}}{\text{no. kg solvent}} \times \frac{k_f}{\Delta T_f}$$

_____ g

f. What is the molecular mass of naphthalene?   _____

g. The molecular formula of naphthalene is $C_{10}H_8$. Is the
result of Part f consistent with this formula?   _____

Name

Section

Molecular Mass Determination by Freezing Point Depression

(Advance Study Assignment)

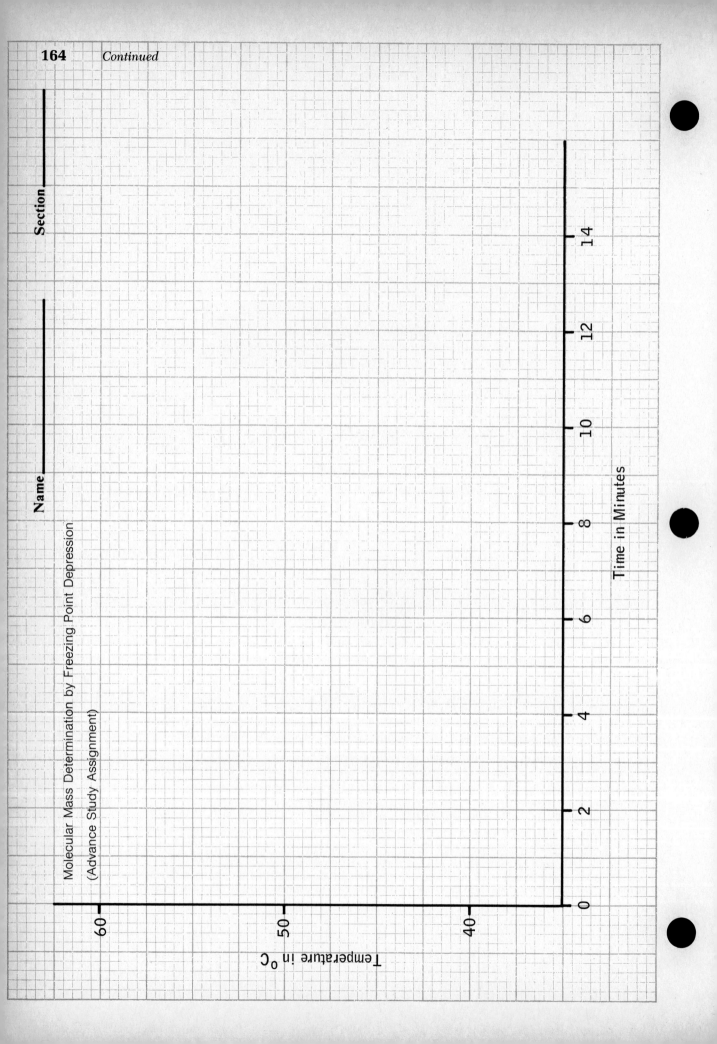

Temperature in °C

60

50

40

Time in Minutes

0   2   4   6   8   10   12   14

EXPERIMENT

20 • The Standardization of a Basic Solution and the Determination of the Equivalent Mass of a Solid Acid

When a solution of a strong acid is mixed with a solution of a strong base, a chemical reaction occurs that can be represented by the following net ionic equation:

$$H^+(aq) + OH^-(aq) \rightarrow H_2O$$

This is called a neutralization reaction, and chemists use it extensively to change the acidic or basic properties of solutions. The equilibrium constant for the reaction is about 10^{14} at room temperature, so that the reaction can be considered to proceed completely to the right, using up whichever of the ions is present in the lesser amount and leaving the solution either acidic or basic, depending on whether H^+ or OH^- ion was in excess.

Since the reaction is essentially quantitative, it can be used to determine the concentrations of acidic or basic solutions. A frequently used procedure involves the titration of an acid with a base. In the titration, a basic solution is added from a buret to a measured volume of acid solution until the number of moles of OH^- ion added is just equal to the number of moles of H^+ ion present in the acid. At that point the volume of basic solution that has been added is measured.

Recalling the definition of the molarity, M_A, of species A, we have

$$M_A = \frac{\text{no. moles of A}}{\text{no. liters of solution, } V} \quad \text{or no. moles of A} = M_A \times V \qquad (1)$$

At the end point of the titration,

$$\text{no. moles } H^+ \text{ originally present} = \text{no. moles } OH^- \text{ added} \qquad (2)$$

So, by Equation 1,

$$M_{H^+} \times V_{\text{acid}} = M_{OH^-} \times V_{\text{base}} \qquad (3)$$

Therefore, if the molarity of either the H^+ or the OH^- ion in its solution is known, the molarity of the other ion can be found from the titration.

The equivalence point or end point in the titration is determined by using a chemical, called an indicator, that changes color at the proper point. The indicators used in acid-base titrations are weak organic acids or bases that change color when they are neutralized. One of the most common indicators is phenolphthalein, which is colorless in acid solutions but becomes red when the pH of the solution becomes 9 or higher.

When a solution of a strong acid is titrated with a solution of a strong base, the pH at the end point will be about 7. At the end point a drop of acid or base added to the solution will change its pH by several pH units, so that phenolphthalein can be used as an indicator in such titrations. If a weak acid is titrated with a strong base, the pH at the equivalence point is somewhat higher than 7, perhaps 8 or 9, and phenolphthalein is still a very satisfactory indicator. If, however, a solution of a weak base such as ammonia is titrated with a strong acid, the pH will be a unit or two less than 7 at the end point, and phenolphthalein will not be as good an indicator for that titration as, for example, methyl red, whose color changes from red to yellow as the pH changes from about 4 to 6. Ordinarily, indicators will be chosen so that their color change occurs at about the pH at the equivalence point of a given acid-base titration.

In this experiment you will determine the molarity of OH^- ion in an NaOH solution by titrating that solution against a standardized solution of HCl. Since in these solutions one mole of acid in solution furnishes one mole of H^+ ion and one mole of base produces one mole of OH^- ion, $M_{HCl} = M_{H^+}$ in the acid solution, and $M_{NaOH} = M_{OH^-}$ in the basic solution. Therefore the titration will allow you to find M_{NaOH} as well as M_{OH^-}.

In the second part of this experiment you will use your standardized NaOH solution to titrate a sample of a pure solid organic acid. By titrating a weighed sample of the unknown acid with your standardized NaOH solution you can, by Equation 2, find the number of moles of H^+ ion that your sample can furnish. From the mass of your sample and the number of moles of H^+ it contains, you can calculate the number of grams of solid acid, GEM, that would contain one mole of H^+ ion.

$$GEM = \frac{\text{no. grams acid}}{\text{no. moles } H^+ \text{ furnished}} \tag{4}$$

The value of GEM for an acid is called the *equivalent mass* of the acid and is equal to the number of grams of acid per mole of available H^+ ion:

$$\text{one equivalent mass acid} \rightarrow \text{one mole } H^+ \text{ ion} \tag{5}$$

The equivalent mass, GEM, of an acid may or may not equal the molar mass of the acid, GMM. The reason is almost, but not quite, obvious. Let us consider three acids, with the molecular formulas HX, H_2Y, and H_3Z:

one mole HX → one mole H^+ Therefore, by Equation 5, GMM = GEM

one mole H_2Y → two moles H^+ ″   GMM = 2 × GEM

one mole H_3Z → three moles H^+ ″   GMM = 3 × GEM

In all cases the molar mass and the equivalent mass are related by simple equations, but in order to find the molar mass from the equivalent mass we need to know the molecular formula of the acid.

EXPERIMENTAL PROCEDURE

Note: This experiment is relatively long unless you know precisely what you are to do. Study the experiment carefully before coming to class, so that you don't have to spend a lot of time finding out what the experiment is all about.

Obtain two burets and a sample of solid unknown acid from the stockroom.

A. Standardization of NaOH Solution. Into a small graduated cylinder draw about 7 ml of the stock 6 M NaOH solution provided in the laboratory and dilute to about 400 ml with distilled water in a 500 ml Florence flask. Stopper the flask tightly and mix the solution thoroughly at intervals over a period of at least 15 minutes before using the solution.

Draw into a clean *dry* 125 ml Erlenmeyer flask about 75 ml of standardized HCl solution (about 0.1 M) from the stock solution on the reagent shelf. This amount should provide all the standard acid you will need; do not waste it. Record the molarity of the HCl.

Prepare for the titration by using the procedure described in Experiment 18. The purpose of this procedure is to make sure that the solution in each buret has the same molarity as it has in the container from which it was poured. Clean the two burets and rinse them with distilled water. Then rinse one buret three times with a few ml of the HCl solution, in each case thoroughly wetting the walls of the buret with the solution and then letting it out through the stopcock. Fill the buret with HCl; open the stopcock momentarily to fill the tip. Proceed to clean and fill the other buret with your NaOH solution in a similar manner. Put the acid buret, A, on the left side of your buret clamp, and the base buret, B, on the right side. Check to see that your burets do not leak and that there are no air bubbles in either buret tip. Read and record the levels in the two burets to 0.02 ml.

Draw about 25 ml of the HCl solution from the buret into a clean 250 ml Erlenmeyer flask; add to the flask about 25 ml of distilled H_2O and 2 or 3 drops of phenolphthalein indicator solution. Place a white sheet of paper under the flask to aid in the detection of any color change. Add the NaOH solution intermittently from its buret to the solution in the flask, noting the pink phenolphthalein color that appears and disappears as the drops hit the liquid and are mixed with it. Swirl the liquid in the flask gently and continuously as you add the NaOH solution. When the pink color begins to persist, slow down the rate of addition of NaOH. In the final stages of the titration add the NaOH drop by drop until the entire solution just turns a pale pink color that will persist for about 30 seconds. If you go past the end point and obtain a red solution, add a few drops of the HCl solution to remove the color, and then add NaOH a drop at a time until the pink color persists. Carefully record the final readings on the HCl and NaOH burets.

To the 250 ml Erlenmeyer flask containing the titrated solution, add about 10 ml more of the standard HCl solution. Titrate this as before with the NaOH to an end point, and carefully record both buret readings once again. To this solution add about 10 ml more HCl and titrate a third time with NaOH.

You have now completed three titrations, with *total* HCl volumes of about 25, 35, and 45 ml. Using Equation 3, calculate the molarity of your base, M_{OH^-}, for each of the three titrations. In each case, use the *total volumes* of acid and base which were added up to that point. At least two of these molarities should agree to within 1%. If they do, proceed to the next part of the experiment. If they do not, repeat these titrations until two calculated molarities do agree.

B. Determination of the Equivalent Mass of an Acid. Weigh the vial containing your solid acid on the analytical balance to ± 0.0001 g. Carefully pour out about half the sample into a clean but not necessarily dry 250 ml Erlenmeyer flask. Weigh the vial accurately. Add about 50 ml of distilled water and 2 or 3 drops of phenolphthalein to the flask. The acid may be relatively insoluble, so don't worry if it doesn't all dissolve.

Fill your NaOH buret with your (now standardized) NaOH solution and read the level accurately.

Titrate the acid solution as before. As the acid is neutralized by the NaOH, it will tend to dissolve in the solution. If your unknown is so insoluble that the phenolphthalein color appears before all the solid dissolves, add 25 ml of ethanol to the solution to increase the solubility. Record the NaOH buret reading at the end point.

Pour the rest of your acid sample into a 250 ml Erlenmeyer flask and weigh the vial accurately. Titrate the acid as before with NaOH solution.

If you go past the end point in these titrations it is possible, although more complicated in calculations, to back-titrate with a little of the standard HCl solution. Measure the volume of HCl used and subtract the number of moles HCl in that volume from the number of moles NaOH used in the titration. The difference will equal the number of moles NaOH used to neutralize the acid sample. For a back titration, at the end point, for volumes in milliliters:

$$\text{no. moles H}^+ \text{ in solid acid} = \frac{M_{NaOH} \times V_{NaOH}}{1000} - \frac{M_{HCl} \times V_{HCl}}{1000} \qquad (5)$$

Name _____ Section _____

DATA: **Standardization of a Basic Solution.**
        Determination of the GEM of an Acid

   A. **Standardization of NaOH Solution**

	Trial 1	Trial 2	Trial 3
Initial reading, HCl buret	_____ ml		
Final reading, HCl buret	_____ ml	_____ ml	_____ ml
Initial reading, NaOH buret	_____ ml		
Final reading, NaOH buret	_____ ml	_____ ml	_____ ml

   B. **Equivalent Mass of Unknown Acid**

Mass of vial plus contents	_____ g
Mass of vial plus contents less Sample 1	_____ g
Mass less Sample 2	_____ g

	Trial 1	Trial 2
Initial reading, NaOH buret	_____ ml	_____ ml
Final reading, NaOH buret	_____ ml	_____ ml
Initial reading, HCl buret	_____ ml	_____ ml
		(if you used back-titration)
Final reading, HCl buret	_____ ml	_____ ml

CALCULATIONS

   A. **Standardization of NaOH Solution**

	Trial 1	Trial 2	Trial 3
Total volume HCl	_____ ml	_____ ml	_____ ml
Total volume NaOH	_____ ml	_____ ml	_____ ml

Continued on following page

Molarity M_{HCl}; of standardized HCl   _____ M

Molarity, M_{H^+}, in standardized HCl   _____ M

By Equation 3,

$$M_{H^+} \times V_{acid} = M_{OH^-} \times V_{base} \text{ or } M_{OH^-} = M_{H^+} \times \frac{V_{HCl}}{V_{NaOH}} \tag{3}$$

Use Equation 3 to find the molarity, M_{OH^-}, of the NaOH solution. Note that the volumes do not need to be converted to liters, since we use the volume ratio.

	Trial 1	Trial 2	Trial 3

M_{OH^-} _____ M   _____ M   _____ M   (should agree within 1%)

The molarity of the NaOH will equal M_{OH^-}, since one mole NaOH $\rightarrow$ one mole OH$^-$.

Average molarity of NaOH solution, M_{NaOH} _____ M

B. Equivalent Mass of Unknown Acid

	Trial 1	Trial 2
Mass of sample	_____ g	_____ g
Volume NaOH used	_____ ml	_____ ml
No. moles NaOH $= \dfrac{V_{NaOH} \times M_{NaOH}}{1000}$	_____	_____
Volume HCl used	_____ ml	_____ ml
No. moles HCl $= \dfrac{V_{HCl} \times M_{HCl}}{1000}$	_____	_____
No. moles H$^+$ in sample (by Eq. 5)	_____	_____
GEM $= \dfrac{\text{no. grams acid}}{\text{no. moles H}^+}$	_____ g	_____ g
Unknown no.	_____	

Name _____ Section _____

ADVANCE STUDY ASSIGNMENT: Equivalent Mass of an Unknown Acid

1. 7.0 ml of 6.0 M NaOH are diluted with water to a volume of 400 ml. We are asked to find the molarity of the resulting solution.
   a. First find out how many moles of NaOH there are in 7.0 ml of 6.0 M NaOH. Use Equation 1. Note that the volume must be in liters.

_____ moles

   b. Since the total number of moles of NaOH is not changed on dilution, the molarity after dilution can also be found by Equation 1, using the final volume of the solution. Calculate that molarity.

_____ M

2. In an acid-base titration, 24.88 ml of an NaOH solution are needed to neutralize 26.43 ml of a 0.1049 M HCl solution. To find the molarity of the NaOH solution, we can use the following procedure:

   a. First note the value of M_{H^+} in the HCl solution.   _____ M

   b. Find M_{OH^-} in the NaOH solution. (Use Eq. 3.)

_____ M

   c. Obtain M_{NaOH} from M_{OH^-}.   _____ M

3. A 0.2349-g sample of an unknown acid requires 33.66 ml of 0.1086 M NaOH for neutralization to a phenolphthalein end point. There are 0.42 ml of 0.1049 M HCl used for back-titration.

   a. How many moles of OH^- are used? How many moles of H^+ from HCl?

_____ moles OH^- _____ moles H^+ from HCl

   b. How many moles of H^+ are there in the solid acid? (Use Eq. 5.)

_____ moles H^+ in solid

Continued on following page   **171**

c. What is the equivalent mass of the unknown acid? (Use Eq. 4.)

_____g

EXPERIMENT

21 • pH, Its Measurement and Applications

One of the more important properties of an aqueous solution is its concentration of hydrogen ion. The H^+ or H_3O^+ ion has great effect on the solubility of many inorganic and organic species, on the nature of complex metallic cations found in solutions, and on the rates of many chemical reactions. It is important that we know how to measure the concentration of hydrogen ion and understand its effect on solution properties.

For convenience the concentration of H^+ ion is frequently expressed as the pH of the solution rather than as molarity. The pH of a solution is defined by the following equation:

$$pH = -\log[H^+] \qquad (1)$$

where the logarithm is taken to the base 10. If $[H^+]$ is 1×10^{-4} moles per liter, the pH of the solution is, by the equation, 4. If the $[H^+]$ is 5×10^{-2} M, the pH is 1.3.

Basic solutions can also be described in terms of pH. In water solutions the following equilibrium relation will always be obeyed:

$$[H^+] \times [OH^-] = K_w = 1 \times 10^{-14} \text{ at } 25°C \qquad (2)$$

In distilled water $[H^+]$ equals $[OH^-]$, so, by Equation 2, $[H^+]$ must be 1×10^{-7} M. Therefore, the pH of distilled water is 7. Solutions in which $[H^+] > [OH^-]$ are said to be acidic and will have a pH < 7; if $[H^+] < [OH^-]$, the solution is basic and its pH > 7. A solution with a pH of 10 will have a $[H^+]$ of 1×10^{-10} M and a $[OH^-]$ of 1×10^{-4} M.

We measure the pH of a solution experimentally in two ways. In the first of these we use a chemical called an indicator, which is sensitive to pH. These substances have colors that change over a relatively short pH range (about 2 pH units) and can, when properly chosen, be used to roughly determine the pH of a solution. Two very common indicators are litmus, usually used on paper, and phenolphthalein, the most common indicator in acid-base titrations. Litmus changes from red to blue as the pH of a solution goes from about 6 to about 8. Phenolphthalein changes from colorless to red as the pH goes from 8 to 10. A given indicator is useful for determining pH only in the region in which it changes color. Indicators are available for measurement of pH in all the important ranges of acidity and basicity. By matching the color of a suitable indicator in a solution of known pH with that in an unknown solution, one can determine the pH of the unknown to within about 0.3 pH units.

The other method for finding pH is with a device called a pH meter. In this device two electrodes, one of which is sensitive to $[H^+]$, are immersed in a solution. The potential between the two electrodes is related to the pH. The pH meter is designed so that the scale will directly furnish the pH of the solution. A pH meter gives much more precise measurement of pH than does a typical indicator, and is ordinarily used when an accurate determination of pH is needed.

Some acids and bases undergo substantial ionization in water, and are called *strong* because of their essentially complete ionization in reasonably dilute solutions. Other acids and bases, because of incomplete ionization (often only about 1 per cent in 0.1 M solution), are called *weak*. Hydrochloric acid, HCl, and sodium hydroxide, NaOH are typical examples of a strong acid and a strong base respectively. Acetic acid, $HC_2H_3O_2$, and ammonia, NH_3, are classic examples of a weak acid and a weak base.

A weak acid will ionize according to the Law of Chemical Equilibrium:

$$HA(aq) \rightleftharpoons H^+(aq) + A^-(aq) \qquad (3)$$

173

At equilibrium,

$$\frac{[H^+][A^-]}{[HA]} = K_a \tag{4}$$

K_a is a constant characteristic of the acid HA; in solutions containing HA, the product of concentrations in the equation will remain constant at equilibrium independent of the manner in which the solution was made up. A similar relation can be written for solutions of a weak base.

The value of the ionization constant K_a for a weak acid can be found experimentally in several ways. One very simple procedure involves very little calculation, is accurate, and does not even require a knowledge of the molarity of the acid. A sample of a weak acid, HA, often a solid, is dissolved in water. The solution is divided into two equal parts. One part of the solution is titrated to a phenolphthalein end point with an NaOH solution, with the HA converted to A^- by the reaction:

$$OH^-(aq) + HA(aq) \rightarrow A^-(aq) + H_2O \tag{5}$$

The number of moles of A^- produced equals the number of moles HA in the other part of the sample. The two solutions are then mixed, and the pH of the resultant solution is obtained. In that solution it is clear that [HA] equals $[A^-]$, so that by (4),

$$[H^+] = K_a \tag{6}$$

By this method the $[H^+]$ obtained from the pH measurement is equal to the value of the ionization constant of the acid.

Salts that can be formed by the reaction of strong acids and bases, such as NaCl, KBr, or $NaNO_3$, ionize completely but do not react with water when in solution. They form neutral solutions with a pH of about 7. When dissolved in water, salts of weak acids or weak bases furnish ions that tend to react to some extent with water, producing molecules of the weak acid or base and liberating some OH^- or H^+ ion to the solution.

If HA is a weak acid, the A^- ion produced when NaA is dissolved in water will react with water to some extent, according to the equation

$$A^-(aq) + H_2O \rightleftharpoons HA(aq) + OH^-(aq) \tag{7}$$

Solutions of sodium acetate, $NaC_2H_3O_2$, the salt formed by reaction of sodium hydroxide with acetic acid, will be slightly *basic* because of the reaction of $C_2H_3O_2^-$ ion with water to produce $HC_2H_3O_2$ and OH^-. Because of the analogous reaction of the NH_4^+ ion with water to form H_3O^+ ion, solutions of ammonium chloride, NH_4Cl, will be slightly *acidic*.

Salts of most transition metal ions are acidic. A solution of $CuSO_4$ or $FeCl_3$ will typically have a pH equal to 5 or lower. The salts are completely ionized in solution. The acidity comes from the fact that the cation is hydrated (e.g., $Cu(H_2O)_4^{2+}$ or $Fe(H_2O)_6^{3+}$). The large + charge on the metal cation attracts electrons from the O—H bonds in water, weakening them and producing some H^+ ions in solution; with $CuSO_4$ solutions the reaction would be

$$Cu(H_2O)_4^{2+}(aq) \rightleftharpoons Cu(H_2O)_3OH^+(aq) + H^+(aq) \tag{8}$$

Some solutions, called buffers, are remarkably resistant to pH changes caused by the addition of an acid or base. These solutions almost always contain both the salt of a weak acid or base and the parent acid or base. The solution used in the previously mentioned procedure for finding K_a of a weak acid is an example of a buffer solution. That solution contained equal amounts of the weak acid HA and the anion A^- present in its salt. If a small amount of strong acid were added to that solution, the H^+ ion would tend to react with the A^- ion present, keeping $[H^+]$ about where it was before the addition. Similarly, a small amount of strong base added to that solution would react with the HA present, producing some A^- ion and water but not appreciably changing $[OH^-]$. If similar amounts of acid or base were added to *water* the pH could easily change by several units.

In this experiment you will first measure the pH of an unknown solution by using acid-base indicators. You will then use a pH meter to find the pH of some typical solutions. Finally, you will measure the dissociation constant of an unknown weak acid by the method described in the discussion. In the process you will prepare a buffer, whose properties will be investigated.

EXPERIMENTAL PROCEDURE

You may work in pairs on the first two parts of this experiment.

A.   Determination of the pH of a Solution by Using Acid-Base Indicators.   Obtain a solution of unknown pH from your laboratory assistant. Put about 1 cm^3 of the solution into a small test tube and then add a drop or two of one of the indicators in Table 21.1. Note the color of the solution you obtain. Repeat the test with each of the indicators in the table. From the information in the table, estimate the pH of the solution roughly. Note that the color of an indicator is most indicative of the pH in the region where the indicator is changing color.

Table 21.1

Indicator	Useful pH Range (Approximate)								Color Change
	0	1	2	3	4	5	6	7	
Methyl violet	←———————→								yellow to violet
Thymol blue		←————————→							red to yellow
Methyl orange				←————————→					red to yellow
Bromphenol blue				←————————→					yellow to blue
Bromcresol green						←————————→			yellow to blue

Note: The arrows indicate the approximate pH range over which the indicator changes color. Outside that range the color remains constant. For example, methyl orange is red from pH 0 to pH 3 and yellow at pH 5 or higher. Between pH 3 and pH 5 the color changes gradually from red to yellow and is orange at about pH 4.

Having established the pH of your solution to within about one pH unit, obtain a known solution of HCl with a pH about equal to that of your unknown. Since HCl is a strong acid and completely ionized, [H$^+$] in 0.1 M HCl is 1×10^{-1} M and the pH is 1.0. Similarly, the pH of 0.001 M HCl is 3.0. Stock solutions of 0.1 M and 0.001 M HCl are available on your lab bench. If you dilute 5 cm^3 of these solutions with 45 cm^3 of distilled water and mix well, you will obtain solutions of pH equal to 2.0 and 4.0, respectively.

Test 1-cm^3 portions of your known HCl solution with the indicator or indicators that were most useful in fixing the pH of your unknown. Compare the colors of your known and unknown solutions. On the basis of the comparison, decide whether the unknown is of higher, or lower pH than the known. Then obtain a second known HCl solution, using the stock solution or a dilution thereof, which differs from the first known by one pH unit and which, along with your first known, should bracket the pH of your unknown. That is, if your first known has a pH of 3 and the unknown is of higher pH, make the second known have a pH of 4; if the unknown had a pH lower than 3, you would make the pH of the second known equal to 2. Treat your second known HCl solution with the useful indicator(s) and compare its color(s) with that of your unknown. If your unknown has a pH between those of your two knowns, estimate its pH by comparing its color with those of the knowns. If it does not, prepare another known HCl solution, continuing until you do get two known solutions which bracket your unknown. For the final estimation of pH, use 5-cm^3 portions of your unknown and of the two known solutions and two drops of indicator in each solution. By using a well-lit white piece of paper as background, and viewing down the three tubes, with the unknown in the middle, you should be able to estimate the pH of the unknown to ± 0.3 pH units.

B.   Measurement of the pH of Some Typical Solutions.   In the rest of the experiment we will be using pH meters to find pH. Your instructor will show you how to operate your meter. The electrodes are fragile, so use due caution when using the pH meter.

Using a 25-ml sample in a 150-ml beaker, measure and record the pH of a 0.1 M solution of each of the following substances:

$$NaCl \quad HC_2H_3O_2 \quad Na_2CO_3$$
$$HCl \quad NH_3 \quad ZnCl_2$$

For each solution write a net ionic equation which explains, qualitatively, or, if possible, quantitatively, why the observed values of pH are reasonable.

C. Determination of the Dissociation Constant of a Weak Acid. In the rest of the experiment each student is to work independently.

Obtain from the stockroom a sample of an unknown acid and a buret. Measure out 100 ml of distilled water into a graduated cylinder and pour it into a clean 250-ml Erlenmeyer flask. Dissolve your acid sample in the water and stir thoroughly.

Pour half the solution into another 250-ml Erlenmeyer flask. Use the solution levels in the two flasks to decide when the volumes of solution in the two flasks are equal. Titrate the acid in one of the flasks to a phenolphthalein end point, using 0.2 M NaOH in the buret. (See Exp. 22; volume readings do not have to be taken here.) This should take less than 50 ml of the NaOH. Add the sodium hydroxide solution slowly while rotating the flask. As the end point approaches, add the solution drop by drop until the solution has a permanent pink color.

Mix the neutralized solution with the acid solution in the other flask and determine the pH of the resulting half-neutralized solution. From the observed pH calculate K_a for the unknown acid.

D. Properties of a Buffer. The solution you prepared in Part C, whose pH you measured, is a buffer. It contains a weak acid, HA, and its conjugate base, A$^-$, in equal concentrations. This solution is much more resistant to pH change than water would be, since the acid, HA, it contains can react with added OH$^-$ ion, and the anion, A$^-$, can react with added H$^+$ ion.

To 25 ml of this buffer solution add 5 drops 0.1 M HCl and mix thoroughly. Measure the pH of the resulting solution. Measure the pH of the distilled water in the laboratory. To a 25-ml sample of distilled water add 5 drops of 0.1 M HCl. After mixing, measure the pH of the solution.

To another 25-ml sample of the buffer, add 5 drops 0.1 M NaOH. Mix well, and measure the pH of the resulting solution. Then add 5 drops 0.1 M NaOH to 25 ml of distilled water and measure the pH.

Write a net ionic equation showing why the buffer solution resists change in pH when H$^+$ ion is added. Write a second equation showing how the buffer reacts to resist pH change when OH$^-$ ion is added.

OBSERVATIONS, CALCULATIONS, and EXPLANATIONS: pH, Its Measurement and Applications

A. Determination of pH Using Acid-Base Indicators

Indicator	Color with Unknown	Color with Known HCl Solution
Methyl violet	_____	_____
Thymol blue	_____	_____
Methyl orange	_____	_____
Bromphenol blue	_____	_____
Bromcresol green	_____	_____

Approx. pH of unknown _____

pH of first known HCl solution _____

Indicator(s) used in final pH determination _____ _____

pH of second known HCl solution _____

pH of third known HCl solution (if needed) _____

Approximate pH of unknown solution ($\pm$ 0.3) _____

Unknown no. _____

B. Measurement of the pH of Some Typical Solutions

Record the pH of the 0.1 M solutions that were tested.

NaCl _____ $HC_2H_3O_2$ _____ Na_2CO_3 _____

HCl _____ NH_3 _____ $ZnCl_2$ _____

Continued on following page   **177**

Write a net ionic equation for the reaction responsible for the observed pH value for each solution.

NaCl _____ $HC_2H_3O_2$ _____

HCl _____ Na_2CO_3 _____

NH_3 _____ $ZnCl_2$ _____

C. Determination of the Dissociation Constant of a Weak Acid

pH of the half-neutralized acid solution _____

$[H^+]$ in the half-neutralized solution _____ M

K_a of the unknown acid _____

Unknown no. _____

D. Properties of an HA-A⁻ Buffer

	Buffer	Buffer + 5 drops 0.1 M HCl	Buffer + 5 drops 0.1 M NaOH
pH	_____	_____	_____

	Water	Water + 5 drops 0.1 M HCl	Water + 5 drops 0.1 M NaOH
pH	_____	_____	_____

Equation for reaction of buffer with H^+ ion: _____

Equation for reaction of buffer with OH^- ion: _____

ADVANCE STUDY ASSIGNMENT: pH, Its Measurement and Applications

1. A 0.5 M solution of an acid was tested with the indicators used in this experiment. The colors observed were as follows:

Methyl violet:	violet	Bromphenol blue:	green
Thymol blue:	yellow	Bromcresol green:	yellow
Methyl orange:	orange		

What is the approximate pH of the solution?   _____

Which indicators would you use to find the pH more precisely?

_____   _____

2. A solution of NaCN has a pH of 10. The CN^- ion is the anion of the weak acid HCN. Write the net ionic equation for the reaction which produces the pH of the NaCN solution (See Eq. 7).

3. The pH of a half-neutralized solution of a weak acid is 4.3. What is the dissociation constant of the weak acid? (See Eq. 6)

K_a = _____

4. A solution with a volume of one liter contains 0.5 mole $HC_2H_3O_2$, 0.5 mole $C_2H_3O_2^-$ ion, and 0.5 mole Na^+ ion. K_a for $HC_2H_3O_2$ is 1.8×10^{-5}.

   a. What is the pH of the solution? (See Eq. 4)   _____

   b. Is the solution a buffer?   _____

   c. Why, or why not?

   d. Write the net ionic equation for the reaction which occurs if a little H^+ ion is added to the solution. (See Eq. 3, and think about it a little bit.)

EXPERIMENT

22 • Determination of the Solubility Product of PbI$_2$

In this experiment you will determine the solubility product of lead iodide, PbI$_2$. Lead iodide is relatively insoluble, having a solubility of less than 0.002 mole per liter at 20°C. The equation for the solution reaction of PbI$_2$ is

$$PbI_2(s) \rightleftarrows Pb^{2+}(aq) + 2I^-(aq) \tag{1}$$

The solubility product expression associated with this reaction is

$$K_{sp} = [Pb^{2+}] [I^-]^2 \tag{2}$$

Equation 2 implies that in any system containing solid PbI$_2$ in equilibrium with its ions, the product of $[Pb^{2+}]$ times $[I^-]^2$ will at a given temperature have a fixed magnitude, independent of how the equilibrium system was initially made up.

In the first part of the experiment, known volumes of standard solutions of Pb(NO$_3$)$_2$ and KI will be mixed in several different proportions. The yellow precipitate of PbI$_2$ formed will be allowed to come to equilibrium with the solution. The value of $[I^-]$ in the solution will be measured experimentally. The $[Pb^{2+}]$ will be calculated from the initial composition of the system, the measured value of $[I^-]$, and the stoichiometric relation between Pb^{2+} and I$^-$ in Equation 1.

By mixing the solutions as we have described, we approach equilibrium by precipitating PbI$_2$ and measuring the concentrations of I$^-$ and Pb^{2+} remaining in the solution. We will also carry out the reaction in the other direction, by first precipitating PbI$_2$, washing it free of excess ions, and then dissolving the solid in the inert salt solution. Under such conditions the concentrations of Pb^{2+} and I$^-$ in the saturated solution will be related by Equation 1, since both ions come from pure PbI$_2$. From the measured value of $[I^-]$ in the saturated solution we can calculate $[Pb^{2+}]$ immediately.

The concentration of I$^-$ ion will be found spectrophotometrically, as in Experiment 17. Although the iodide ion is not colored, it is relatively easily oxidized to I$_2$, which is brown in water solution. Our procedure will be to separate the solid PbI$_2$ from the solution and then to oxidize the I$^-$ in solution with potassium nitrite, KNO$_2$, under slightly acidic conditions, where the conversion to I$_2$ is quantitative. Although the concentration of I$_2$ will be rather low in the solutions you will prepare, the absorption of light by I$_2$ in the vicinity of 525 nm is sufficiently intense to make accurate analyses possible.

In all of the solutions prepared, potassium nitrate, KNO$_3$ (note the distinction between KNO$_2$ and KNO$_3$!), will be present as an inert salt. This salt serves to keep the ionic strength of the solution essentially constant at 0.2 M and promotes the formation of well-defined crystalline precipitates of PbI$_2$.

EXPERIMENTAL PROCEDURE

From the stock solutions that are available, measure out about 35 ml of 0.012 M Pb(NO$_3$)$_2$ in 0.20 M KNO$_3$ into a small beaker. To a second small beaker add 30 ml 0.030 M KI in 0.20 M KNO$_3$ and, to a third, add 10 ml 0.20 M KNO$_3$. Use the labeled graduated cylinders next to each of the reagent bottles for measuring out these solutions. Use these reagent solutions in your experiment.

Label five regular test tubes 1 to 5, either with labels or by noting their positions in your test tube rack. Into the first four tubes pipet 5.0 ml of 0.012 M Pb(NO$_3$)$_2$ in KNO$_3$. Then, to test tube 1, add 2.0 ml 0.030 M KI in KNO$_3$. Add 3.0, 4.0, and 5.0 ml of that same solution to test tubes

2, 3, and 4, respectively. Add enough 0.20 M KNO_3 to the first three tubes to make the total volume 10.0 ml in each tube. The composition of the final mixture in each tube is summarized in Table 22.1.

TABLE 22.1   VOLUMES OF REAGENTS USED IN PRECIPITATING PbI_2 (Ml)

Test Tube	0.012 M $Pb(NO_3)_2$	0.030 M KI	0.20 M KNO_3
1	5.0	2.0	3.0
2	5.0	3.0	2.0
3	5.0	4.0	1.0
4	5.0	5.0	0.0
5	10.0	10.0	0.0

Stopper each test tube and shake thoroughly at intervals of several minutes while you are proceeding with the next part of the experiment.

In the fifth test tube mix about 10 ml of 0.012 M $Pb(NO_3)_2$ in KNO_3 with 10 ml of 0.03 M KI in KNO_3. Shake the mixture vigorously for a minute or so. Let the solid settle for a few minutes and then decant and discard three-fourths of the supernatant solution. Transfer the solid PbI_2 and the rest of the solution to a small test tube and centrifuge. Discard the liquid, retaining the solid precipitate. Add 3 ml 0.20 M KNO_3 and shake to wash the solid free of excess Pb^{2+} or I^-. Centrifuge again, and discard the liquid. By this procedure you should now have prepared a small sample of essentially pure PbI_2 in a little KNO_3 solution. Add 0.20 M KNO_3 to the solid until the tube is about three-fourths full. Shake well at several one minute intervals to saturate the solution with PbI_2.

In this experiment it is essential that the volumes of reagents used to make up the mixtures in test tubes 1 to 4 be measured accurately. It is also essential that all five mixtures be shaken thoroughly so that equilibrium can be established. Insufficient shaking of the first four test tubes will result in not enough PbI_2 precipitating to reach true equilibrium; if the small test tube is not shaken sufficiently, not enough PbI_2 will dissolve to attain equilibrium.

When each of the mixtures has been shaken for at least 15 minutes, let the tubes stand for three to four minutes to let the solid settle. Pour the supernatant liquid in test tube 1 into a small dry test tube until it is three-fourths full and centrifuge for about three minutes to settle the solid PbI_2. Pour the liquid into another small dry test tube; if there are any solid particles or yellow color remaining in the liquid, centrifuge again. When you have a clear liquid, dip a small piece of clean, dry paper towel into the liquid to remove floating PbI_2 particles from the surface. Pipet 3.0 ml of 0.02 M KNO_2, potassium NITRITE (*not* KNO_3, potassium nitrate), into a clean, dry spectrophotometer tube and add 2 drops 6 M HCl. Then, using a medicine dropper, add enough of the clear centrifuged solution (about 3 ml) to fill the spectrophotometer tube just to the level indicated on the tube. Shake gently to mix the reagents and then measure the absorbance of the solution as directed by your instructor. The calibration curve or equation which is provided will allow you to determine directly the concentration of I^- ion that was in equilibrium with PbI_2. Use the same procedure to analyze the solutions in test tubes 2 through 5, completing each analysis before you proceed to the next.

DATA AND CALCULATIONS: Determination of the Solubility Product of PbI$_2$

From the experimental data we obtain [I$^-$] directly. To obtain K$_{sp}$ for PbI$_2$ we must calculate [Pb^{2+}] in each equilibrium system. This is most easily done by constructing an equilibrium table. We first find the initial amounts of I$^-$ and Pb^{2+} ions in each system from the way the mixtures were made up. Knowing [I$^-$] and the formula of lead iodide allows us to calculate [Pb^{2+}]. K$_{sp}$ then follows directly. The calculations are similar to those in Experiment 14. Some suggestions regarding the calculations are on the following page. Use them if you need to.

$$PbI_2(s) \rightleftharpoons Pb^{2+}(aq) + 2I^-(aq) \qquad K_{sp} = [Pb^{2+}]\,[I^-]^2$$

DATA

Test tube no.	1	2	3	4	Saturated solution of PbI$_2$ 5
ml 0.012 M Pb(NO$_3$)$_2$	_____	_____	_____	_____	
ml 0.03 M KI	_____	_____	_____	_____	
ml 0.20 M KNO$_3$	_____	_____	_____	_____	
Total volume in ml	_____	_____	_____	_____	
Absorbance of solution	_____	_____	_____	_____	_____
[I$^-$] in moles/liter at equilibrium	_____	_____	_____	_____	_____

CALCULATIONS

1. Initial no. moles Pb^{2+} ____ × 10^{-5} ____ × 10^{-5} ____ × 10^{-5} ____ × 10^{-5}

2. Initial no. moles I$^-$ ____ × 10^{-5} ____ × 10^{-5} ____ × 10^{-5} ____ × 10^{-5}

3. No. moles I$^-$ at equilibrium   ____ × 10^{-5} ____ × 10^{-5} ____ × 10^{-5} ____ × 10^{-5}

4. No. moles I$^-$ precipitated   ____ × 10^{-5} ____ × 10^{-5} ____ × 10^{-5} ____ × 10^{-5}

Continued on following page   **183**

5. No. moles Pb^{2+}
   precipitated   _____ $\times 10^{-5}$ _____ $\times 10^{-5}$ _____ $\times 10^{-5}$ _____ $\times 10^{-5}$

6. No. moles Pb^{2+} at
   equilibrium   _____ $\times 10^{-5}$ _____ $\times 10^{-5}$ _____ $\times 10^{-5}$ _____ $\times 10^{-5}$

7. $[Pb^{2+}]$ at equilibrium   _____ _____ _____ _____ _____

8. K_{sp} PbI_2 _____ _____ _____ _____ _____

Suggestions Regarding Calculations

In test tubes 1 to 4, calculations must be based on the amounts of reagents which were used. In test tube 5, Sections 1–6 do not apply; the solution is saturated with PbI_2, so initial amounts or reagents are not relevant. Several of the calculations depend upon the definition of molarity. For species A,

$$M_A = \frac{\text{no. moles of A}}{\text{volume of soln}} \quad \text{or no. moles of A} = M_A \times \text{volume (liters)} \tag{6}$$

1. The initial number of moles of Pb^{2+} is equal to the amount in the $Pb(NO_3)_2$ solution that was used. What is $[Pb^{2+}]$ in that solution? _____M. What volume was used? _____ ml; _____ liters. Use Equation 6 to find the initial number of moles of Pb^{2+}. Since the volumes used were equal, each tube (1–4) contains the same number of moles of Pb^{2+}.

2. What is $[I^-]$ in the KI solution? _____M. What volumes were used? _____ liters. Use Equation 6 to find the initial numbers of moles of I^- present. Since the solution volumes in test tubes 1 to 4 differ, the numbers of moles of I^- differ.

3. Here we need to find the number of moles of I^- in each solution at equilibrium. What is the total volume of each solution? _____ ml; _____ liters (same for tubes 1 to 4). From $[I^-]$ as measured at equilibrium, and the solution volume, find the number of moles of I^- in each solution at equilibrium, again using Equation 6.

4. The number of moles of I^- in solution at equilibrium is given by the equation.

no. moles I^- in soln = initial no. moles I^- − no. moles I^- precipitated

From the no. of moles of I^- you started with (Section 2) and the number remaining in solution (Section 3), you can easily obtain the number of moles of I^- that precipitated.

5. The number of moles of Pb^{2+} that precipitated is related to the number of moles of I^- that precipitated, since they both come down in the PbI_2. What is the relationship? _____ _____ Calculate the number of moles Pb^{2+} that precipitated from the number of moles of I^- that precipitated.

6. The number of moles Pb^{2+} in solution at equilibrium must equal the number initially present minus the number that precipitated. Make that calculation for tubes 1 to 4.

Continued on following page

7. What was the total volume of solution for each of the test tubes 1 to 4? _____ ml;

_____ liters. Use Equation 6 to find [Pb^{2+}] in the solution in each test tube.
In the solution in test tube 5, the Pb^{2+} and the I^- both came from dissolving PbI_2, so the concentra-

tions of those two ions are related. What is the relationship? _____ Calculate [Pb^{2+}] from [I^-] in tube 5.

8. From [Pb^{2+}] and [I^-] at equilibrium, find K_{sp} for PbI_2 for each of the solutions in test tubes 1
to 5, using Equation 2.

ADVANCE STUDY ASSIGNMENT: Determination of the Solubility Product of PbI_2

1. State in words the meaning of the solubility product equation for PbI_2:

$$K_{sp} = [Pb^{2+}][I^-]^2$$

2. When 5.0 ml of 0.012 M $Pb(NO_3)_2$ are mixed with 5.0 ml of 0.030 M KI, a yellow precipitate of $PbI_2(s)$ forms.

   a. How many moles of Pb^{2+} are initially present? (See Section 1, Suggestions Regarding Calculations.)

   _____ moles

   b. How many moles of I^- are originally present (Section 2)?

   _____ moles

   c. In a colorimeter the equilibrium solution is analyzed for I^-, and its concentration is found to be 8×10^{-3} mole/liter. How many moles of I^- are present in the solution (10 ml) (Section 3)?

   _____ moles

   d. How many moles of I^- precipitated (Section 4)?

   _____ moles

   e. How many moles of Pb^{2+} precipitated (Section 5)?

   _____ moles

   f. How many moles of Pb^{2+} are left in solution (Section 6)?

   _____ moles

   g. What is the concentration of Pb^{2+} in the equilibrium solution (Section 7)?

   _____ moles/liter

   h. Find a value for K_{sp} of PbI_2 from these data (Section 8).

   _____

187

EXPERIMENT

23 • Relative Stabilities of Complex Ions and Precipitates Prepared from Solutions of Copper(II)

In aqueous solution, typical cations, particularly those produced from atoms of the transition metals, do not exist as free ions but rather consist of the metal ion in combination with some water molecules. Such cations are called complex ions. The water molecules, usually 2, 4, or 6 in number, are bound chemically to the metallic cation, but often rather loosely, with the electrons in the chemical bonds being furnished by one of the unshared electron pairs from the oxygen atoms in the H_2O molecules. Copper ion in aqueous solution may exist as $Cu(H_2O)_4^{2+}$, with the water molecules arranged in a square around the metal ion at the center.

If a hydrated cation such as $Cu(H_2O)_4^{2+}$ is mixed with other species that can, like water, form coordinate covalent bonds with Cu^{2+}, those species, called ligands, may displace one or more H_2O molecules and form other complex ions containing the new ligands. For instance, NH_3, a reasonably good coordinating species, may replace H_2O from the hydrated copper ion, $Cu(H_2O)_4^{2+}$, to form $Cu(H_2O)_3NH_3^{2+}$, $Cu(H_2O)_2(NH_3)_2^{2+}$, $Cu(H_2O)(NH_3)_3^{2+}$, or $Cu(NH_3)_4^{2+}$. At moderate concentrations of NH_3, essentially all the H_2O molecules around the copper ion are replaced by NH_3 molecules, forming the copper ammonia complex ion.

Coordinating ligands differ in their tendency to form bonds with metallic cations, so that in a solution containing a given cation and several possible ligands, an equilibrium will develop in which most of the cations are coordinated with those ligands with which they form the most stable bonds. There are many kinds of ligands, but they all share the common property that they possess an unshared pair of electrons which they can donate to form a coordinate covalent bond with a metal ion. In addition to H_2O and NH_3, other uncharged coordinating species include CO and ethylenediamine; some common anions that can form complexes include OH^-, Cl^-, CN^-, SCN^-, and $S_2O_3^{2-}$.

As you know, when solutions containing metallic cations are mixed with other solutions containing ions, precipitates are sometimes formed. When a solution of 0.1 M copper nitrate is mixed with a little 1 M NH_3 solution, a precipitate forms and then dissolves in excess ammonia. The formation of the precipitate helps us to understand what is occurring as NH_3 is added. The precipitate is hydrous copper hydroxide, formed by reaction of the hydrated copper ion with the small amount of hydroxide ion present in the NH_3 solution. The fact that this reaction occurs means that even at very low OH^- ion concentration $Cu(OH)_2(H_2O)_2(s)$ is a more stable species than $Cu(H_2O)_4^{2+}$ ion.

Addition of more NH_3 causes the solid to redissolve. The copper species then in solution cannot be the hydrated copper ion. (Why?) It must be some other complex ion, and is, indeed, the $Cu(NH_3)_4^{2+}$ ion. The implication of this reaction is that the $Cu(NH_3)_4^{2+}$ ion is also more stable in NH_3 solution than is the hydrated copper ion. To deduce in addition that the copper ammonia complex ion is also more stable in general than $Cu(OH)_2(H_2O)_2(s)$ is not warranted, since under the conditions in the solution $[NH_3]$ is much larger than $[OH^-]$, and given a higher concentration of hydroxide ion, the solid hydrous copper hydroxide might possibly precipitate even in the presence of substantial concentrations of NH_3.

To resolve this question, you might proceed to add a little 1 M NaOH solution to the solution containing the $Cu(NH_3)_4^{2+}$ ion. If you do this you will find that $Cu(OH)_2(H_2O)_2(s)$ does indeed precipitate. We can conclude from these observations that $Cu(OH)_2(H_2O)_2(s)$ is more stable than $Cu(NH_3)_4^{2+}$ in solutions in which the ligand concentrations (OH^- and NH_3) are roughly equal.

The copper species that will be present in a system depends, as we have just seen, on the conditions in the system. We cannot say in general that one species will be more stable than another; the stability of a given species depends in large measure on the kinds and concentrations of other species that are also present with it.

Another way of looking at the matter of stability is through equilibrium theory. Each of the copper species we have mentioned can be formed in a reaction between the hydrated copper ion and a complexing or precipitating ligand; each reaction will have an associated equilibrium constant, which we might call a formation constant for that species. The pertinent formation reactions and their constants for the copper species we have been considering are listed here:

$$Cu(H_2O)_4{}^{2+}(aq) + 4\,NH_3(aq) \rightleftharpoons Cu(NH_3)_4{}^{2+}(aq) + 4\,H_2O \qquad K_1 = 5 \times 10^{12} \qquad (1)$$

$$Cu(H_2O)_4{}^{2+}(aq) + 2\,OH^-(aq) \rightleftharpoons Cu(OH)_2(H_2O)_2(s) + 2\,H_2O \qquad K_2 = 2 \times 10^{19} \qquad (2)$$

The expressions for the formation constants for these reactions do not involve $[H_2O]$ terms, which are essentially constant in aqueous systems and are included in the magnitude of K in each case. The large size of each formation constant indicates that the tendency for the hydrated copper ion to react with the ligands listed is very high.

In terms of these data, let us compare the stability of the $Cu(NH_3)_4{}^{2+}$ complex ion with that of solid $Cu(OH)_2(H_2O)_2$. This is most readily done by considering the reaction:

$$Cu(NH_3)_4{}^{2+}(aq) + 2\,OH^-(aq) + 2\,H_2O \rightleftharpoons Cu(OH)_2(H_2O)_2(s) + 4\,NH_3(aq) \qquad (3)$$

We can find the value of the equilibrium constant for this reaction by noting that it is the sum of Reaction 2 and the reverse of Reaction 1. By the Law of Multiple Equilibrium, K for Reaction 3 is given by the equation

$$K = \frac{K_2}{K_1} = \frac{2 \times 10^{19}}{5 \times 10^{12}} = 4 \times 10^6 = \frac{[NH_3]^4}{[Cu(NH_3)_4{}^{2+}][OH^-]^2} \qquad (4)$$

From the expression in Equation 4 we can calculate that in a solution in which the NH_3 and OH^- ion concentrations are both about 1 M,

$$[Cu(NH_3)_4{}^{2+}] = \frac{1}{4 \times 10^6} = 2.5 \times 10^{-7}\ M \qquad (5)$$

Since the concentration of the copper ammonia complex ion is very, very low, any copper(II) in the system will exist as the solid hydroxide. In other words, the solid hydroxide is more stable under such conditions than the ammonia complex ion. But that is exactly what we observed when we treated the hydrated copper ion with ammonia and then with an equivalent amount of hydroxide ion.

Starting now from the experimental behavior of the copper ion, we can conclude that since the solid hydroxide is the species that exists when copper ion is exposed to equal concentrations of ammonia and hydroxide ion, the hydroxide is more stable under those conditions, *and* the equilibrium constant for the formation of the hydroxide is larger than the constant for the formation of the ammonia complex. By determining, then, which species is present when a cation is in the presence of equal ligand concentrations, we can speak meaningfully of stability under such conditions and can rank the formation constants for the possible complex ions, and indeed for precipitates, in order of their increasing magnitudes.

In this experiment you will carry out formation reactions for a group of complex ions and precipitates involving the Cu^{2+} ion. You can make these species by mixing a solution of $Cu(NO_3)_2$ with solutions containing NH_3 or anions, which may form either precipitates or complex ions by reaction with $Cu(H_2O)_4{}^{2+}$, the cation present in aqueous solutions of copper(II) nitrate. By examining whether the precipitates or complex ions formed by the reaction of hydrated copper(II) ion with a given species can, on addition of a second ligand, be dissolved or transformed to another species, you will be able to rank the relative stabilities of the precipitates and complex ions made from Cu^{2+} with respect to one another, and thus rank the equilibrium formation constants for each species in order of increasing magnitude. The species to be reacted with Cu^{2+} ion in aqueous solution are NH_3, Cl^-, OH^-, CO_3^{2-}, $C_2O_4^{2-}$, S^{2-}, NO_2^-, and PO_4^{3-}. In each case the test for relative stability will be made in the presence of essentially equal concentrations of the two ligands. When you have completed your ranking of the known species you will test an unknown species and incorporate it into your list.

EXPERIMENTAL PROCEDURE

Obtain from the stockroom an unknown and enough small test tubes so that you have a total of eight.

Add about 2 ml 0.1 M $Cu(NO_3)_2$ solution to each of the test tubes.

To one of the test tubes add about 2 ml 1 M NH_3, drop by drop. Note whether a precipitate forms initially, and if it dissolves in excess NH_3. Shake the tube sideways to mix the reagents. In the NH_3-NH_3 space in the table (data page), write a P in the upper left hand corner if a precipitate is formed initially. In the rest of the space, write the formula and the color of the species which was present with an excess of NH_3. If a solution is present, the copper ion will be in a complex. Cu(II) will always have a coordination number of 4, so the formula with NH_3 would be $Cu(NH_3)_4^{2+}$. If a precipitate is present, it will be neutral, and in the case of NH_3 it would be a hydroxide with the formula $Cu(OH)_2$ (there should in principle be two H_2O molecules in the formula, but they are usually omitted). Add two ml 1 M NH_3 to the rest of the test tubes.

Now you will test the stability of the species present in excess NH_3 relative to those which might be present with other precipitating or coordinating species. Add, drop by drop, 2 ml of a 1 M solution of each of the anions in the horizontal row of the table to the test tubes you have just prepared, one solution to a test tube. Note any changes that occur in the appropriate spaces in the table. A change in color or the formation of a precipitate implies that a reaction has occurred between the added ligand or precipitating anion and the species originally present. As before, put a P in the upper left hand corner if a precipitate initially forms on addition of the anion solution. In the rest of the space, write the formula and color of the species present when the anion is in excess. That is the species which is stable in the presence of equal concentrations of NH_3 and the added anion. Again, in complex ions, Cu(II) will normally be attached to four ligands; copper(II) precipitates will be neutral. If a new species forms on addition of the second reagent, its formula should be given. If no change occurs, the original species is more stable, so put its formula in that space.

Repeat the above series of experiments, using 1 M Cl^- as the species originally added to the $Cu(NO_3)_2$ solution. In each case record the color of any precipitates or solutions formed on addition of the reagents in the horizontal row, and the formula of the species stable when an excess of both Cl^- ion and the added species is present in the solution. Since these reactions are reversible, it is not necessary to retest Cl^- with NH_3, since the same results would be obtained as when the NH_3 solution was tested with Cl^- solution.

Repeat the series of experiments for each of the anions in the vertical row in the table, omitting those tests where decisions as to relative stabilities are already clear. Where both ligands produce precipitates it will be helpful to check the effect of the addition of the other ligand to those precipitates. When complete, your table should have at least 36 entries.

Examine your table and decide on the relative stabilities of all species you observed to be present in all the spaces of the table. There should be eight such species; rank them as best you can in order of increasing stability. There is only one correct ranking, and you should be prepared to defend your choices. Although we did not in general prepare the species by direct reaction of $Cu(H_2O)_4^{2+}$ with the added ligand or precipitating anion, the equilibrium formation constants for those species for the direct reactions will have magnitudes that increase in the same order as the relative stabilities of the species you have established.

When you are satisfied that your ranking order is correct, carry out the necessary tests on your unknown to determine its proper position in the list. Your unknown may be one of the Cu(II) species you have already observed, or it may be a different species, present in excess of its ligand or precipitating anion. If your unknown contains a precipitate, shake it well before using a portion of it to make a test.

Name _____ Section _____

DATA AND OBSERVATIONS: Relative Stabilities of Complex Ions and Precipitates Containing Cu(II)

Table of Observations

	NH$_3$	Cl$^-$	OH$^-$	CO$_3^{2-}$	C$_2$O$_4^{2-}$	S^{2-}	NO$_2^-$	PO$_4^{3-}$	Unknown
PO$_4^{3-}$									
NO$_2^-$									
S^{2-}									
C$_2$O$_4^{2-}$									
CO$_3^{2-}$									
OH$^-$									
Cl$^-$									
NH$_3$									

Continued on following page   **193**

Determination of Relative Stabilities

In each row of the table you can compare the stabilities of species involving the reagent in the horizontal row with those of the species containing the reagent initially added. In the first row of the table, the copper(II)–NH$_3$ species can be seen to be more stable than some of the species obtained by addition of the other reagents, and less stable than others. Examining each row, make a list of all the complex ions and precipitates you have in the table in order of increasing stability and formation constant.

Reasons

Lowest _____   _____

_____   _____

_____   _____

_____   _____

_____   _____

_____   _____

Highest _____   _____

Stability of Unknown

Indicate the position your unknown would occupy in the above list.

Reasons:

Unknown no. _____

ADVANCE STUDY ASSIGNMENT: Stabilities of Complex Ions and Precipitates of Cu(II)

1. In testing the relative stabilities of Cu(II) species a student adds 2 ml 1 M NH_3 to 2 ml 0.1 M $Cu(NO_3)_2$. He observes that a blue precipitate initially forms, but that in excess NH_3 the precipitate dissolves and the solution turns dark blue. Addition of 2 ml 1 M NaOH to the dark blue solution results in the formation of a blue precipitate.

     a. What is the formula of Cu(II) species in the dark blue solution?

     b. What is the formula of the blue precipitate present after addition of 1 M NaOH?

     c. Which species is more stable in equal concentrations of NH_3 and OH^-, the one in Part a or the one in Part b?

                                                    _____

2. Given the following two reactions and their equilibrium constants:

$$Cu(H_2O)_4{}^{2+}(aq) + 4\,NH_3(aq) \rightleftharpoons Cu(NH_3)_4{}^{2+}(aq) + 4\,H_2O \qquad K_1 = 5 \times 10^{12}$$

$$Cu(H_2O)_4{}^{2+}(aq) + CO_3{}^{2-} \rightleftharpoons CuCO_3(s) + 4\,H_2O \qquad K_2 = 7 \times 10^9$$

     a. Evaluate the equilibrium constant for the reaction

$$Cu(NH_3)_4{}^{2+}(aq) + 2\,CO_3{}^{2-} \rightleftharpoons CuCO_3(s) + 4\,NH_3$$

                                                    _____

     b. Given the value of K in part a and the relationship between stability and the formation constants, K_1 and K_2, which Cu(II) species would be stable in a solution in which $[NH_3] = [CO_3{}^{2-}] = 1$ M?

                                                    _____

EXPERIMENT

24 • Synthesis of Some Coordination Compounds

Some of the most interesting research in inorganic chemistry has involved the preparation and study of the properties of those substances known as coordination compounds. These compounds, sometimes called complexes, are typically salts that contain *complex ions*. A complex ion is an ion that contains a central metal ion to which are bonded small polar molecules or simple ions; the bonding in the complex ion is through coordinate covalent bonds, which ordinarily are relatively weak.

In this experiment we shall be concerned with the synthesis of two coordination compounds:

$$\text{A. } [Cu(NH_3)_4]\,SO_4 \cdot H_2O \qquad \text{B. } [Co(NH_3)_6]\,Cl_3$$

The complex ions in these substances are enclosed in brackets to indicate those species which are bonded to the central ion. In this experiment you will prepare the complex ions by making use of reactions in which substituting *ligands*, or coordinating species, replace other ligands on the central ion. The reactions will usually be carried out in water solution, in which the metallic cation will initially be present in the simple hydrated form; addition of a reagent containing a complexing ligand will result in an exchange reaction of the sort

$$Cu(H_2O)_4{}^{2+}(aq) + 4\,NH_3(aq) \rightleftharpoons Cu(NH_3)_4{}^{2+}(aq) + 4\,H_2O \qquad (1)$$

In many reactions involving complex ion formation the rate of reaction is very rapid, so that the thermodynamically stable form of the ion is the one produced. Such reactions obey the law of chemical equilibrium and can thus be readily controlled as to direction by a change in the reaction conditions. Reaction 1 proceeds readily to the right in the presence of NH_3 in moderate concentrations. However, by decreasing the NH_3 concentration, for example by the addition of acid to the system, we can easily regenerate the hydrated copper cation. Complex ions that undergo very fast exchange reactions, such as those in Reaction 1, are called *labile*.

Most but by no means all complex ions are labile. Some complex ions, including one to be studied in this experiment, will exchange ligands only slowly. For such species, called *inert* or *nonlabile*, the complex ion produced in a substitution reaction may be the one which is kinetically rather than thermodynamically favored. Alteration of reaction conditions, perhaps by the addition of a catalyst, may change the relative rates of formation of possible complex products and so change the complex ion produced in the reaction. In your experiment you will use a catalyst, activated charcoal, in the preparation of the complex ion in compound B. In the absence of the catalyst the complex ion that would form would be $Co(NH_3)_5H_2O^{3+}$. The Co(III) present in these species is stabilized in complex ions containing NH_3 ligands.

Many complex ions are highly colored, both in solution and in the solid salt. An easy way to determine whether a complex ion is labile is to note whether a color change occurs in a solution containing the ion when a reagent is added that reacts with the ligands in that complex ion. We will use this procedure to establish the labile or nonlabile character of the complex ions you prepare.

EXPERIMENTAL PROCEDURE

A. Preparation of [Cu(NH$_3$)$_4$] SO$_4$ · H$_2$O

$$Cu(H_2O)_4^{2+}(aq) + SO_4^{2-}(aq) + 4\,NH_3(aq) \rightarrow [Cu(NH_3)_4]\,SO_4 \cdot H_2O(s) + 3\,H_2O$$

Weigh out 7.0 g of CuSO$_4$ · 5 H$_2$O on a triple-beam or other rough balance. Weigh the solid either on a piece of paper or in a small beaker and not directly on the balance pan. Transfer the solid copper sulfate to a 125 ml Erlenmeyer flask and add 15 ml of water. Heat the flask to dissolve the solid, then cool to room temperature.

Carry out the remaining steps under the hood. Add 15 M NH$_3$ solution, a few ml at a time, swirling the flask to mix the reagents, until the first precipitate has completely dissolved. All the copper should now be present in solution as the complex ion Cu(NH$_3$)$_4^{2+}$.

The sulfate salt of this complex cation can be precipitated by the addition of a liquid, such as ethyl alcohol, in which Cu(NH$_3$)$_4$SO$_4$ · H$_2$O is insoluble. Add 10 ml of 95% ethyl alcohol to the solution; this should result in the formation of a deep blue precipitate of Cu(NH$_3$)$_4$SO$_4$ · H$_2$O. Filter the solution through a Buchner funnel, using suction. Wash the solid in the funnel by adding two 5 ml portions of 95% ethanol. Dry the solid by pressing it between two pieces of filter paper. Put the crystals on a piece of weighed filter paper and let them dry further in the air. When they are thoroughly dry, weigh them on the paper to 0.1 g.

B. Synthesis of Hexamminecobalt(III) Chloride, [Co(NH$_3$)$_6$] Cl$_3$

$$2\,Co(H_2O)_6^{2+}(aq) + 6\,Cl^-(aq) + 12\,NH_3(aq) + H_2O_2(aq)$$
$$\rightarrow 2\,[Co(NH_3)_6]\,Cl_3(s) + 12\,H_2O + 2\,OH^-(aq)$$

Obtain a 5-ml sample of cobalt(II) chloride solution (containing 1.0 g CoCl$_2$). Pour this solution into a 150-ml beaker and add 1.2 g NH$_4$Cl, 0.1 g activated charcoal (Norite), and 5 ml distilled water. Heat the solution to the boiling point, stirring to complete the solution of the NH$_4$Cl.

Cool the beaker in an ice bath, and slowly add 5 ml 15 M NH$_3$. Stir until well mixed and then slowly, drop by drop, with stirring, add 5 ml of 10% hydrogen peroxide, H$_2$O$_2$.

When the bubbling has stopped, place the 150-ml beaker in a 600-ml beaker containing 100 ml water at about 60°C. Leave the beaker in the water bath for 20 minutes, holding the temperature of the bath at 60 ± 5°C by judicious use of your Bunsen burner. Stir the mixture occasionally.

Remove the 150-ml beaker from the water bath and return it to the ice bath. Cool for about 5 minutes, stirring to promote crystallization of the crude product. Set up a Büchner funnel, and, with suction, filter the mixture through the filter paper in the funnel. Discard the filtrate.

Scrape the solid from the filter paper into a 150-ml beaker. Add 50 ml H$_2$O and 2 ml 12 M HCl. Heat the mixture to boiling, with stirring. Using distilled water, rinse out the suction flask used with the Büchner funnel. Holding the beaker with a pair of tongs, filter the hot mixture through the funnel, with suction on. The charcoal is removed at this point and should remain on the filter paper; the filtrate should be a golden yellow.

Transfer the filtrate to a 150-ml beaker and add 8 ml 12 M HCl. Place the beaker in an ice bath and stir for several minutes to promote formation of the golden crystals of [Co(NH$_3$)$_6$] Cl$_3$. Separate the crystals by filtration through the Büchner funnel. Turn the suction off, add 10 ml 95% ethanol, and let stand for about 10 seconds. Turn the suction on and pull off the ethanol, which should carry with it much of the water and HCl remaining on the crystals. Draw air through the crystals for several minutes. Transfer the product to a piece of weighed filter paper. Let it dry for several more minutes and weigh it on the paper to 0.1 g.

C. Relative Lability of Complex Ions. Dissolve a small portion (about 0.5 g) of the complexes from Parts A and B, $[Cu(NH_3)_4]SO_4 \cdot H_2O$ and $[Co(NH_3)_6]Cl_3$, in a few ml of water. Note the color and observe the effect of the addition of a few drops of 12 M HCl.

On completion of Part C, show your products from Parts A and B to your instructor for evaluation.

CAUTION: In these experiments you will be using 15 M NH_3, 12 M HCl, and a 10% solution of H_2O_2. These are caustic reagents, so use them carefully. Avoid getting them on your skin or clothing, and don't breathe the vapors of NH_3 or HCl. If you do come into contact with any of these reagents, wash them off thoroughly with water.

DATA AND OBSERVATIONS: Synthesis of Some Coordination Compounds

	$[Cu(NH_3)_4]\,SO_4 \cdot H_2O$	$[Co(NH_3)_6]\,Cl_3$
Mass of filter paper	_____ g	_____ g
Mass of filter paper plus product	_____ g	_____ g
Mass of product	_____ g	_____ g
Theoretical yield	_____ g	_____ g
Percentage yield	_____ %	_____ %

Observations Regarding Lability of Complexes (Part C)

	Color of solid	Color in H_2O solution	Color on addition of HCl
$Cu(NH_3)_4{}^{2+}$	_____	_____	_____
$Co(NH_3)_6{}^{3+}$	_____	_____	_____

Conclusions

What evidence, if any, do you have that the formulas of the two compounds you prepared are correct, either as to the nature of the atoms or groups present or as to the number of the groups present? Consider the nature of the possible cations, anions, and ligands in the compounds.

Name _____ Section _____

ADVANCE STUDY ASSIGNMENT: Synthesis of Some Coordination Compounds

1. Calculate the theoretical yields of the compounds to be prepared in this experiment. The metal ion in both cases is the limiting reagent. Hint: Find the number of moles of Cu(II) in the sample of $CuSO_4 \cdot 5 H_2O$ that you used. That will equal the number of moles of $[Cu(NH_3)_4] SO_4 \cdot H_2O$ that could theoretically be prepared. Proceed in a similar way for the synthesis involving Co(II).

A.

_____ g

B.

_____ g

2. How could you establish that the metal ion is the limiting reagent?

3. How would you formally name the compound $[Cu(NH_3)_4] SO_4 \cdot H_2O$ prepared in Part A? the $[Co(NH_3)_6] Cl_3$ in Part B?

EXPERIMENT

25 • Determination of an Equivalent Mass by Electrolysis

In Experiment 20 we determined the equivalent mass of an acid by titration of a known mass of the acid with a standardized solution of NaOH. There we defined an equivalent mass of acid to be the amount of acid in grams that could furnish one mole of H^+ ion.

Experimentally we find that the equivalent mass of an element can be related in a fundamental way to the chemical effects observed in that phenomenon known as *electrolysis*. As you know, some liquids, because they contain ions, will conduct an electric current. If the two terminals on a storage battery, or any other source of D.C. voltage, are connected through metal electrodes to a conducting liquid, an electric current will pass through the liquid and chemical reactions will occur at the two metal electrodes; in this process electrolysis is said to occur, and the liquid is said to be electrolyzed.

At the electrode connected to the *negative* pole of the battery, a *reduction* reaction will invariably be observed. In this reaction electrons will usually be accepted by one of the species present in the liquid, which, in the experiment we shall be doing, will be an aqueous solution. The species reduced will ordinarily be a metallic cation or the H^+ ion or possibly water itself; the reaction which is actually observed will be the one that occurs with the least expenditure of electrical energy, and will depend on the composition of the solution. In the electrolysis cell we shall study, the reduction reaction of interest will occur in a slightly acidic medium; hydrogen gas will be produced by the reduction of hydrogen ion:

$$2\,H^+(aq) + 2\,e^- \rightarrow H_2(g) \tag{1}$$

In this reduction reaction, which will occur at the negative pole, or *cathode*, of the cell, for every H^+ ion reduced *one* electron will be required, and for every molecule of H_2 formed, *two* electrons will be needed.

Ordinarily in chemistry we deal not with individual ions or molecules but rather with moles of substances. In terms of moles, we can say that, by Equation 1,

> The reduction of one mole of H^+ ion requires one mole of electrons
> The production of one mole of $H_2(g)$ requires two moles of electrons

A mole of electrons is a fundamental amount of electricity in the same way that a mole of pure substance is a fundamental unit of matter, at least from a chemical point of view. A mole of electrons is called a *faraday*, after Michael Faraday, who discovered the basic laws of electrolysis. The amount of a species which will react with a *mole* of *electrons*, or *one faraday*, is equal to the *equivalent mass* of that species. Since one faraday will reduce one mole of H^+ ion, we say that the equivalent mass of hydrogen is 1.008 grams, equal to the mass of one mole of H^+ ion (or one-half mole of $H_2(g)$). To form one mole of $H_2(g)$ one would have to pass two faradays through the electrolysis cell.

In the electrolysis experiment we will perform we will measure the volume of hydrogen gas produced under known conditions of temperature and pressure. By using the Ideal Gas Law we will be able to calculate how many moles of H_2 were formed, and hence how many faradays of electricity passed through the cell.

At the positive pole of an electrolysis cell (the metal electrode that is connected to the + terminal of the battery), an *oxidation* reaction will occur, in which some species will give up electrons. This reaction, which takes place at the *anode* in the cell, may involve again an ionic or neutral species in the solution or the metallic electrode itself. In the cell that you will be studying, the pertinent oxidation reaction will be that in which a metal under study will participate:

$$M(s) \rightarrow M^{n+}(aq) + ne^- \tag{2}$$

During the course of the electrolysis the atoms in the metal electrode will be converted to metallic cations and will go into the solution. The mass of the metal electrode will decrease, depending on the amount of electricity passing through the cell and the nature of the metal. In order to oxidize one mole, or one molar mass, of the metal, it would take n faradays, where n is the charge on the cation which is formed. By definition, one faraday of electricity would cause one equivalent mass, GEM, of metal to go into solution. The molar mass, GAM, and the equivalent mass of the metal are related by the equation:

$$GAM = GEM \times n \tag{3}$$

In an electrolysis experiment, since n is not determined independently, it is not possible to find the molar mass of a metal. It is possible, however, to find equivalent masses of many metals, and that will be our main purpose.

The general method we will use is implied by the discussion. We will oxidize a sample of an unknown metal at the positive pole of an electrolysis cell, weighing the metal before and after the electrolysis and so determining its loss in mass. We will use the same amount of electricity, the same number of electrons, to reduce hydrogen ion at the negative pole of the electrolysis cell. From the volume of H_2 gas which is produced under known conditions we can calculate the number of moles of H_2 formed, and hence the number of faradays which passed through the cell. The equivalent mass of the metal is then calculated as the amount of metal which would be oxidized if one faraday were used. In an optional part of the experiment, your instructor may tell you the nature of the metal you used. Using Equation 3, it will be possible to determine the charge on the metallic cations that were produced during electrolysis.

EXPERIMENTAL PROCEDURE

Obtain from the stockroom a buret and a sample of a metal unknown. Weigh the metal sample on the analytical balance to 0.0001 g.

Set up the electrolysis apparatus as indicated in Figure 25.1. There should be about 100 ml 0.1 M $HC_2H_3O_2$ in 0.5 M Na_2SO_4 in the beaker with the gas buret. This will serve as the conducting solution. Immerse the end of the buret in the solution and attach a length of rubber tubing to its upper end. Open the stopcock on the buret and, with suction, carefully draw the acid up to the top of the graduations. Close the stopcock. Insert the bare coiled end of the heavy copper wire up into the end of the buret; all but the coil end of the wire should be covered with watertight insulation. Check the solution level after a few minutes to make sure the stopcock does not leak. Record the level.

The metal unknown will serve as the anode in the electrolysis cell. Connect the metal to the + pole of the power source with an alligator clip and immerse the metal but not the clip in the conducting solution. The copper electrode will be the cathode in the cell. Connect that electrode to the − pole of the power source. Hydrogen gas should immediately begin to bubble from the copper cathode. Collect the gas until about 50 ml have been produced. At that point, stop the electrolysis by disconnecting the copper electrode from the power source. Record the level of the liquid in the buret. Measure and record the temperature of the solution in the beaker and the barometric pressure in the laboratory. (In some cases a cloudiness may develop in the solution during the electrolysis. This is caused by the formation of a metal hydroxide, and will have no adverse effect on the experiment.)

Raise the buret, and discard the conducting solution in the beaker. Rinse the beaker with water, and pour in 100 ml of fresh conducting solution. Repeat the electrolysis, generating about

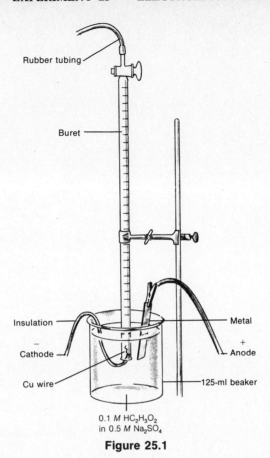

Figure 25.1

50 ml of H_2 and recording the initial and final liquid levels in the buret. Take the alligator clip off the metal anode and wash the anode with 0.1 M $HC_2H_3O_2$, acetic acid. Rub off any loose adhering coating. Rinse the electrode in water and then dry it by immersing it in acetone and letting the acetone evaporate. Weigh the metal electrode to the nearest 0.0001 g.

Name _____ Section _____

DATA AND CALCULATIONS: Determination of an Equivalent Mass by Electrolysis

Mass of metal anode   _____ g

Mass of anode after electrolysis   _____ g

Initial buret reading   _____ ml

Buret reading after first electrolysis   _____ ml

Buret reading after refilling   _____ ml

Buret reading after second electrolysis   _____ ml

Barometric pressure   _____ mm Hg

Temperature t _____ °C

Vapor pressure of H_2O at t _____ mm Hg

Total volume of H_2 produced, V _____ ml

Temperature T _____ K

Pressure exerted by dry H_2: $P = P_{Bar} - VP_{H_2O}$
(ignore any pressure effect due to liquid levels in buret)   _____ mm Hg

No. moles H_2 produced, n
(use Ideal Gas Law, $PV = nRT$)   _____ moles

No. of faradays passed   _____

Loss in mass by anode   _____ g

Equivalent mass of metal $\left(\text{GEM} = \dfrac{\text{no. g lost}}{\text{no. faradays passed}}\right)$ _____ g

Unknown metal number   _____

Optional: Nature of metal _____ GAM _____ g

Charge n on cation _____ (Eq 3)

ADVANCE STUDY ASSIGNMENT: Determination of an Equivalent
Mass by Electrolysis

1. In an electrolysis cell similar to the one employed in this experiment, a student observed that his unknown metal anode lost 0.233 g while a total volume of 94.50 ml of H_2 was being produced. The temperature in the laboratory was 25°C and the barometric pressure was 748 mm Hg. At 25°C the vapor pressure of water is 23.8 mm Hg. To find the equivalent mass of his metal, he filled in the blanks below. Fill in the blanks as he did.

$P_{H_2} = P_{Bar} - VP_{H_2O} =$ _____ mm Hg = _____ atm

$V_{H_2} =$ _____ ml = _____ liters

$T =$ _____ K

$n_{H_2} =$ _____ moles $n_{H_2} = \dfrac{PV}{RT}$

1 mole H_2 requires passage of _____ faradays

No. of faradays passed = _____

Loss of mass of metal anode = _____ grams

No. grams of metal lost per faraday passed $= \dfrac{\text{no. grams lost}}{\text{no. faradays passed}} =$ _____ g = GEM

The student was told that his metal anode was made of copper.

GAM Cu = _____ g   The charge n on the Cu ion is therefore _____ (Eq 3)

EXPERIMENT

26 • Voltaic Cell Measurements

Many chemical reactions can be classified as oxidation-reduction reactions, since they involve the oxidation of one species and the reduction of another. Such reactions can conveniently be considered as the result of two half-reactions, one of oxidation and the other reduction. In the case of the oxidation-reduction reaction

$$Zn(s) + Pb^{2+}(aq) \rightarrow Zn^{2+}(aq) + Pb(s)$$

which would occur if a piece of metallic zinc were put into a solution of lead nitrate, the two reactions would be

$$Zn(s) \rightarrow Zn^{2+}(aq) + 2\ e^- \qquad \text{oxidation}$$

$$2\ e^- + Pb^{2+}(aq) \rightarrow Pb(s) \qquad \text{reduction}$$

The tendency for an oxidation-reduction reaction to occur can be measured if the two reactions are made to occur in separate regions connected by a barrier that is porous to ion movement. An apparatus called a *voltaic cell* in which this reaction might be carried out under this condition is shown in Figure 26.1.

If we connect a voltmeter between the two electrodes we will find that there is a voltage, or potential, between them. The magnitude of the potential is a direct measure of the driving force or thermodynamic tendency of the spontaneous oxidation-reduction reaction to occur.

If we study several oxidation-reduction reactions we find that the voltage of each associated voltaic cell can be considered to be the sum of a potential for the oxidation reaction and a potential for the reduction reaction. In the Zn, $Zn^{2+} \| Pb^{2+}$, Pb cell we have been discussing, for example,

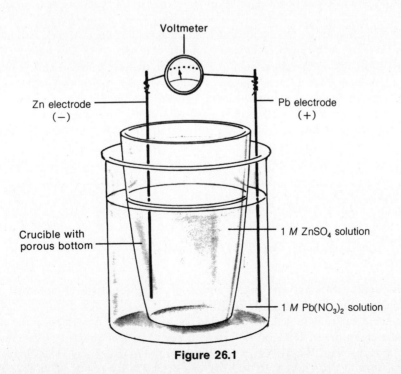

Figure 26.1

213

$$E_{\text{cell}} = E_{\text{Zn, Zn}^{2+} \text{ oxi lation reaction}} + E_{\text{Pb}^{2+}, \text{ Pb reduction reaction}} \tag{1}$$

By convention, the negative electrode in a voltaic cell is taken to be the one from which electrons are emitted (i.e., where oxidation occurs). The negative electrode is the one which is connected to the minus pole of the voltmeter when the voltage is measured.

Since any cell potential is the sum of two electrode potentials, it is not possible, by measuring cell potentials, to determine individual absolute electrode potentials. However, if a value of potential is arbitrarily assigned to one electrode reaction, then other electrode potentials can be given definite values, based on the assigned value. The usual procedure is to assign a value of 0.0000 volts to the standard potential for the electrode reaction

$$2\,H^+(aq) + 2\,e^- \rightarrow H_2(g); \quad E_{H^+,\,H_2\,\text{red}} = 0.0000\ V$$

For the Zn, Zn$^{2+}\|$H$^+$, H$_2$ cell, the measured potential is 0.76 V, and the zinc electrode is negative. Zinc metal is therefore oxidized, and the cell reaction must be

$$Zn(s) + 2\,H^+(aq) \rightarrow Zn^{2+}(aq) + H_2(g); \quad E_{\text{cell}} = 0.76\ V$$

Given this information, one can readily find the potential for the oxidation of Zn to Zn^{2+}.

$$E_{\text{cell}} = E_{\text{Zn, Zn}^{2+} \text{ oxid}} + E_{H^+,\,H_2\,\text{red}}$$

$$0.76\ V = E_{\text{Zn, Zn}^{2+}\text{ oxid}} + 0.00\ V; \quad E_{\text{Zn, Zn}^{2+}\text{ oxid}} = +0.76\ V$$

If the potential for a half-reaction is known, the potential for the reverse reaction can be obtained by changing the sign. For example:

$$\textit{if } E_{\text{Zn, Zn}^{2+}\text{ oxid}} = +0.76 \text{ volts}, \quad \textit{then } E_{\text{Zn}^{2+},\,\text{Zn red}} = -0.76 \text{ volts}$$

$$\textit{if } E_{\text{Pb}^{2+},\,\text{Pb red}} = +Y \text{ volts}, \quad \textit{then } E_{\text{Pb, Pb}^{2+}\text{ oxid}} = -Y \text{ volts}$$

In the first part of this experiment you will measure the voltages of several different cells. By arbitrarily assigning the potential of a particular half-reaction to be 0.00 V, you will then be able to calculate the potentials corresponding to all of the various half reactions that occurred in your cells.

In our discussion so far we have not considered the possible effects of such system variables as temperature, potential at the liquid-liquid junction, size of metal electrodes, and concentrations of solute species. Although temperature and liquid junctions do have a definite effect on cell potentials, taking account of their influence involves rather complex thermodynamic concepts and is usually not of concern in any elementary course. The size of a metal electrode has no appreciable effect on electrode potential, although it does relate directly to the capacity of the cell to produce useful electrical energy. In this experiment we will operate the cells so that they deliver essentially no energy but exert their maximum potentials.

The effect of solute ion concentrations is important and can be described relatively easily. For the cell reaction at 25°C:

$$aA(s) + bB^+(aq) \rightarrow cC(s) + dD^{2+}(aq)$$

$$E_{\text{cell}} = E^0_{\text{cell}} - \frac{0.06}{n} \log \frac{(\text{conc. D}^{2+})^d}{(\text{conc. B}^+)^b} \tag{2}$$

where E^0_{cell} is a constant for a given reaction and is called the standard cell potential, and n is the number of electrons in either electrode reaction.

By Equation 2 you can see that the measured cell potential, E_{cell}, will equal the standard cell potential if the molarities of D^{2+} and B$^+$ are both unity, or, if d equals b, if the molarities are simply equal to each other. We will carry out our experiments under such conditions that the cell potentials you observe will be very close to the standard potentials given in the tables in your chemistry text.

Considering the $Cu,Cu^{2+}\|Ag^+,Ag$ cell as a specific example, the observed cell reaction would be

$$Cu(s) + 2\,Ag^+(aq) \rightarrow Cu^{2+}(aq) + 2\,Ag(s)$$

For this cell, Equation 2 takes the form

$$E_{cell} = E_{cell}^0 - \frac{0.06}{2} \log \frac{conc.\ Cu^{2+}}{(conc.\ Ag^+)^2} \tag{3}$$

In the equation n is 2 because in the cell reaction, two electrons are transferred in each of the two half-reactions. E^0 would be the cell potential when the copper and silver salt solutions are both 1 M, since then the logarithm term is equal to zero.

If we decrease the Cu^{2+} concentration, keeping that of Ag^+ at 1 M, the potential of the cell will go up by about 0.03 volts for every factor of ten by which we decrease conc. Cu^{2+}. Ordinarily it is not convenient to change concentrations of an ion by several orders of magnitude, so in general, concentration effects in cells are relatively small. However, if we should add a complexing or precipitating species to the copper salt solution, the value of conc. Cu^{2+} would drop drastically, and the voltage change would be appreciable. In the experiment we will illustrate this effect by using NH_3 to complex the Cu^{2+}. Using Equation 3, we can actually calculate conc. Cu^{2+} in the solution of its complex ion.

In an analogous experiment we will determine the solubility product of AgCl. In this case we will surround the Ag electrode in a $Cu,Cu^{2+}\|Ag^+,Ag$ cell with a solution of known Cl^- ion concentration which is saturated with AgCl. From the measured cell potential, we can use Equation 3 to calculate the very small value of conc. Ag^+ in the chloride-containing solution. Knowing the concentrations of Ag^+ and Cl^- in a solution in equilibrium with AgCl(s) allows us to find K_{sp} for AgCl.

EXPERIMENTAL PROCEDURE

You may work in pairs in this experiment.

A. Cell Potentials. In this experiment you will be working with these seven electrode systems:

$$Ag^+,\ Ag(s) \qquad Br_2(l),\ Br^-,\ Pt$$
$$Cu^{2+},\ Cu(s) \qquad Cl_2(g,\ 1\ atm),\ Cl^-,\ Pt$$
$$Fe^{3+},\ Fe^{2+},\ Pt \qquad I_2(s),\ I^-,\ Pt$$
$$Zn^{2+},\ Zn(s)$$

Your purpose will be to measure enough voltaic cell potentials to allow you to determine the electrode potential of each electrode by comparing it with an arbitrarily chosen electrode potential.

Using the apparatus shown in Figure 26.1, set up a voltaic cell involving any two of the electrodes in the list. About 10 ml of each solution should be enough for making the cell. The solute ion concentrations may be assumed to be one molar and all other species may be assumed to be at unit activity, so that the potentials of the cells you set up will be essentially the standard potentials. Measure the cell potential and record it along with which electrode has negative polarity.

In a similar manner set up and measure the cell voltage and polarities of other cells, sufficient in number to include all of the electrode systems on the list at least once. Do not combine the silver electrode system with any of the halogen electrode systems, since a precipitate will form; any other combinations may be used. The data from this part of the experiment should be entered in the first three columns of the table in Part A.1 of your report.

B.   Effect of Concentration on Cell Potentials

1.   COMPLEX ION FORMATION.   Set up the $Cu,Cu^{2+}\|Ag^{+},Ag$ cell, using 10 ml of the $CuSO_4$ solution in the crucible and 10 ml of $AgNO_3$ in the beaker. Measure the potential of the cell. While the potential is being measured, add 10 ml of 6 M NH_3 to the $CuSO_4$ solution, stirring carefully with your stirring rod. Measure the potential when it becomes steady.

2.   DETERMINATION OF THE SOLUBILITY PRODUCT OF AgCl. Remove the crucible from the cell you have just studied and discard the $Cu(NH_3)_4{}^{2+}$. Clean the crucible by drawing a little 6 M NH_3 through it, using the adapter and suction flask. Then draw some distilled water through it. Reassemble the Cu-Ag cell, this time using the beaker for the Cu-$CuSO_4$ electrode system. Immerse the Ag electrode in the crucible in 1 M KCl; add a drop of $AgNO_3$ solution to form a little AgCl, so that an equilibrium between Ag^+ and Cl^- can be established. Measure the potential of this cell, noting which electrode is negative. In this case conc. Ag^+ will be very low, which will decrease the potential of the cell to such an extent that its polarity may change from that observed previously.

DATA AND CALCULATIONS: Voltaic Cell Measurements

A.1.   Cell Potentials

Electrode systems used in cell	Cell potential, E^0_{cell} (volts)	Negative electrode	Oxidation reaction	$E^0_{oxidation}$ in volts	Reduction reaction	$E^0_{reduction}$ in volts
1.						
2.						
3.						
4.						
5.						
6.						
7.						

CALCULATIONS

A. Noting that oxidation occurs at the *negative* pole in a cell, write the oxidation reaction in each of the cells. The other electrode system must undergo reduction; write the reduction reaction which occurs in each cell.

B. Assume that $E^0_{Ag^+, Ag} = 0.00$ volts (whether in reduction or oxidation). Enter that value in the table for all of the silver electrode systems you used in your cells. Since $E^0_{cell} = E^0_{oxidation} + E^0_{reduction}$, you can calculate E^0 values for all the electrode systems in which the Ag, Ag^+ system was involved. Enter those values in the table.

C. Using the values and relations in B and taking advantage of the fact that for any given electrode system, $E^0_{oxidation} = -E^0_{reduction}$, complete the table of E^0 values. The best way to do this is to use one of the E^0 values you found in B in another cell with that electrode system. That potential, along with E^0_{cell}, will allow you to find the potential of the other electrode. Continue this process with other cells until all the electrode potentials have been determined.

Continued on following page

A.2. Table of Electrode Potentials

In Table A.1, you should have a value for E^0_{red} or E^0_{oxid} for each of the electrode systems you have studied. Remembering that for any electrode system, $E^0_{red} = -E^0_{oxid}$, you can find the value for E^0_{red} for each system. List those potentials in the left column of the table below in order of decreasing value.

$E^0_{reduction}$ ($E^0_{Ag^+, Ag} = 0.00$ volts)	Electrode reaction in reduction	$E^0_{reduction}$ ($E^0_{H^+, H_2} = 0.00$ volts)
_____	_____	_____
_____	_____	_____
_____	_____	_____
_____	_____	_____
_____	_____	_____
_____	_____	_____
_____	_____	_____

The electrode potentials you have determined are based on $E^0_{Ag^+, Ag} = 0.00$ volts. The usual assumption is that $E^0_{H^+, H_2} = 0.00$ volts, under which conditions $E^0_{Ag^+, Ag\ red} = 0.80$ volts. Convert from one base to the other by adding 0.80 volts to each of the electrode potentials and enter these values in the third column of the table.

Why are the values of E^0_{red} on the two bases related to each other in such a simple way?

B. Effect of Concentration on Cell Potentials

1. Complex ion formation:

Potential, E^0_{cell}, before addition of 6 M NH_3 _____ volts

Potential, E_{cell}, after $Cu(NH_3)_4{}^{2+}$ formed _____ volts

Given Equation 3

$$E_{cell} = E^0_{cell} - \frac{0.06}{2} \log \frac{\text{conc. } Cu^{2+}}{(\text{conc. } Ag^+)^2} \tag{3}$$

Continued on following page

calculate the residual concentration of free Cu^{2+} ion in equilibrium with $Cu(NH_3)_4^{2+}$ in the solution in the crucible. Take conc. Ag^+ to be 1 M.

conc. Cu^{2+} = _____ M

2. Solubility product of AgCl:

Potential, E^0_{cell}, of the Cu, $Cu^{2+} \| Ag^+$, Ag cell (from B.1)

_____V Negative electrode _____

Potential, E_{cell}, with 1 M KCl present   _____V Negative electrode _____

    Using Equation 3, calculate (conc. Ag^+) in the cell, where it is in equilibrium with 1 M Cl^- ion. (E_{cell} in Equation 3 is the *negative* of the measured value if the polarity is not the same as in the standard cell.) Take conc. Cu^{2+} to be 1 M.

conc. Ag^+ = _____ M

    Since Ag^+ and Cl^- in the cell are in equilibrium with AgCl, we can find K_{sp} for AgCl from the concentration of Ag^+ and Cl^-, as they exist in the cell. Formulate the expression for K_{sp} for AgCl, and determine its value.

K_{sp} = _____

ADVANCE STUDY ASSIGNMENT: Voltaic Cell Measurements

1. A student measures the potential of a cell made up with 1 M $CuSO_4$ in one solution and 1 M $AgNO_3$ in the other. There is a Cu electrode in the $CuSO_4$ and an Ag electrode in the $AgNO_3$, and the cell is set up as in Figure 26.1. She finds that the potential, or voltage, of the cell, E^0_{cell}, is 0.45 volts, and that the Cu electrode is negative.

   a. At which electrode is oxidation occurring?   _____

   b. Write the equation for the oxidation reaction.

   c. Write the equation for the reduction reaction.

   d. If the potential of the silver, silver ion electrode, $E^0_{Ag^+,Ag}$ is taken to be 0.000 volts in oxidation or reduction, what is the value of the potential for the oxidation reaction, $E^0_{Cu,Cu^{2+} oxid}$? $E^0_{cell} = E^0_{oxid} + E^0_{red}$.

   _____ volts

   e. If $E^0_{Ag^+,Ag\ red}$ equals 0.80 volts, as in standard tables of electrode potentials, what is the value of the potential of the oxidation reaction of copper, $E^0_{Cu,Cu^{2+} oxid}$?

   _____ volts

   f. Write the net ionic equation for the spontaneous reaction that occurs in the cell which the student studied.

   g. The student adds 6 M NH_3 to the $CuSO_4$ solution until the Cu^{2+} ion is essentially all converted to $Cu(NH_3)_4^{2+}$ ion. The voltage of the cell, E_{cell}, goes up to 0.90 volts and the Cu electrode is still negative. Find the residual concentration of Cu^{2+} ion in the cell. Use Eq. 3.

   _____ M

EXPERIMENT

27 • Preparation of Copper(I) Chloride

Oxidation-reduction reactions are, like precipitation reactions, often used in the preparation of inorganic substances. In this experiment we will employ a series of such reactions to prepare one of the less commonly encountered salts of copper, copper(I) chloride. Most copper compounds contain copper(II), but copper(I) is present in a few slightly soluble or complex copper salts. The process of synthesis of CuCl we will use begins by dissolving copper metal in nitric acid:

$$Cu(s) + 4 H^+(aq) + 2 NO_3^-(aq) \rightarrow Cu^{2+}(aq) + 2 NO_2(g) + 2 H_2O \tag{1}$$

The solution obtained is treated with sodium carbonate in excess, which neutralizes the remaining acid with evolution of CO_2 and precipitates Cu(II) as the carbonate:

$$2 H^+(aq) + CO_3^{2-}(aq) \rightleftharpoons (H_2CO_3)(aq) \rightleftharpoons CO_2(g) + H_2O \tag{2}$$

$$Cu^{2+}(aq) + CO_3^{2-}(aq) \rightleftharpoons CuCO_3(s) \tag{3}$$

The $CuCO_3$ will be purified by filtration and washing and dissolved in hydrochloric acid. Copper metal added to the highly acidic solution then reduces the Cu(II) to Cu(I) and is itself oxidized to Cu(I) in a disproportionation reaction. In the presence of excess chloride, the copper will be present as a $CuCl_4^{3-}$ complex ion. Addition of this solution to water destroys the complex, and white CuCl precipitates.

$$CuCO_3(s) + 2 H^+(aq) + 4 Cl^-(aq) \rightarrow CuCl_4^{2-}(aq) + CO_2(g) + H_2O \tag{4}$$

$$CuCl_4^{2-}(aq) + Cu(s) + 4 Cl^-(aq) \rightarrow 2 CuCl_4^{3-}(aq) \tag{5}$$

$$CuCl_4^{3-}(aq) \xrightarrow{H_2O} CuCl(s) + 3 Cl^-(aq) \tag{6}$$

Since CuCl is readily oxidized, due care must be taken to minimize its exposure to air during its preparation and while it is being dried.

EXPERIMENTAL PROCEDURE

WEAR YOUR SAFETY GLASSES WHILE PERFORMING THIS EXPERIMENT

Obtain a 1 g sample of copper metal, a Buchner funnel, and a filter flask from the stockroom. Weigh the copper metal on the top loading or triple beam balance to 0.1 g.

Put the metal in a 150 ml beaker and *under a hood* add 5 ml 15 M HNO_3. *CAUTION:* Caustic reagent. Brown NO_2 gas will be evolved and an acidic blue solution of $Cu(NO_3)_2$ produced. If it is necessary, you may warm the beaker gently with a Bunsen burner to dissolve all of the copper. When all of the copper is in solution, add 50 ml of water to the solution and allow it to cool.

Weigh out about 5 grams of sodium carbonate in a small beaker on a rough balance. Add small amounts of the Na_2CO_3 to the solution with your spatula, adding the solid as necessary when the evolution of CO_2 subsides. Stir the solution to expose it to the solid. When the acid is neutralized, a blue-green precipitate of $CuCO_3$ will begin to form. At that point, add the rest of the Na_2CO_3, stirring the mixture well to ensure complete precipitation of the copper carbonate.

Transfer the precipitate to the Buchner funnel and use suction to remove the excess liquid.

Use your rubber policeman and a spray from your wash bottle to make a complete transfer of the solid. Wash the precipitate well with distilled water with suction on, then let it remain on the filter paper with suction on for a minute or two.

Remove the filter paper from the funnel and transfer the solid $CuCO_3$ to the 150 ml beaker. Add 10 ml water then 25 ml 6 M HCl slowly to the solid, stirring continuously. When the $CuCO_3$ has all dissolved, add 1.5 g Cu foil cut in small pieces to the beaker and cover it with a watch glass.

Heat the mixture in the beaker to the boiling point and keep it at that temperature, just simmering, for 30 to 40 minutes. It may be that the dark-colored solution which forms will clear to a yellow color before that time is up, and if it does, you may stop heating and proceed with the next step.

While the mixture is heating, put 150 ml distilled water in a 400 ml beaker and put the beaker in an ice bath. Cover the beaker with a watch glass. After you have heated the acidic $Cu-CuCl_2$ mixture for 40 minutes or as soon as it turns light-colored, carefully decant the hot liquid into the beaker of water, taking care not to transfer any of the excess Cu metal to the beaker. White crystals of CuCl should form. Continue to cool the beaker in the ice bath to promote crystallization and to increase the yield of solid.

Cool 25 ml distilled water, to which you have added 5 drops 6 M HCl, in an ice bath. Put 20 ml acetone into a small beaker. Filter the crystals of CuCl in the Buchner funnel using suction. Swirl the beaker to aid in transferring the solid to the funnel. Just as the last of the liquid is being pulled through, wash the CuCl with one third of the acidified cold water. Rinse the last of the CuCl into the funnel with another portion of the water and use the final third to rewash the solid. Turn off suction and add one half of the acetone to the funnel; wait about ten seconds and turn on the suction. Repeat this operation with the other half of the acetone. Draw air through the sample for a few minutes to dry it. If you have properly washed the solid, it will be pure white; if the moist compound is allowed to come into contact with air, it will tend to turn pale green, due to oxidation of Cu(I) to Cu(II). Weigh the CuCl in a previously weighed beaker to 0.1 g. Show your sample to your instructor for evaluation.

Name _____ Section _____

DATA AND RESULTS: Preparation of CuCl

Mass of Cu sample   _____ g

Mass of beaker   _____ g

Mass of beaker plus CuCl   _____ g

Mass of CuCl prepared   _____ g

Theoretical yield   _____ g

Percentage yield   _____ %

ADVANCED STUDY ASSIGNMENT: Preparation of Copper(I) Chloride

1.   The Cu^{2+} ions in this experiment are produced from the reaction of 1.0 g of copper foil with excess nitric acid. How many moles of Cu^{2+} are produced?

_____ moles Cu^{2+}

2.   Why isn't hydrochloric acid used in a direct reaction with copper wire to prepare the $CuCl_2$ solution?

3.   How many grams of metallic copper are required to react with the number of moles of Cu^{2+} calculated in (1) to form the CuCl? The overall reaction can be taken to be:
$Cu^{2+}(ag) + 2 Cl^-(ag) + Cu(s) \rightarrow 2 CuCl(s)$

_____ g Cu

4.   What is the maximum mass of CuCl that can be prepared from the reaction sequence of this experiment using 1.0 g of Cu foil to prepare the Cu^{2+} solution?

_____ g CuCl

28 • Determination of the Half-life of a Radioactive Isotope

Some atomic nuclei are radioactive; that is to say, they can spontaneously undergo reactions in which they converted to different nuclei, which may or may not belong to different elements. For example $^{20}_{9}F$ nuclei may spontaneously emit electrons in a reaction in which $^{20}_{10}Ne$ nuclei are produced:

$$^{20}_{9}F \rightarrow ^{20}_{10}Ne + ^{0}_{-1}e$$

Reactions such as this differ from ordinary chemical reactions in several ways. Typically they produce electrons, He^{2+} nuclei, γ-rays, or some combination of these, all having very high associated energies. The reactions ordinarily involve relatively few atoms, and we usually study them by measuring the high energy particles or radiations that are emitted. Neither the temperature nor the state of chemical combination of the substances containing the active nuclei affect appreciably either the rate at which the reaction occurs or the energies of the emitted species.

The rate at which given radioactive nuclei decay depends only on the kind and number of nuclei present. In any fixed time interval Δt a certain fraction of the active nuclei will undergo reaction according to the equation

$$\frac{\Delta n}{n} = -k\Delta t \tag{1}$$

where n is the number of active nuclei in the sample and Δn is the change in the number of active nuclei (or minus the number undergoing decay) in the time interval Δt. Equation 1 can be rewritten to read

$$\frac{\Delta n}{\Delta t} = -kn \tag{2}$$

which tells us that the rate of radioactive decay, $\Delta n/\Delta t$, is proportional to the number of active nuclei present. The proportionality constant k is called the rate constant for the decay, and the minus sign indicates that the number of active nuclei in the sample decreases with time.

By the methods of the calculus, Equation 2 can be integrated to produce an equation that allows us to calculate the number of active nuclei in a sample at any given time, given the initial number of nuclei n_0 and the associated decay constant:

$$\log_{10}n = \log_{10}n_0 - \frac{kt}{2.3} \tag{3}$$

The time t required for half of the active nuclei in a sample to decay is called the half-life of the nuclei, $t_{1/2}$. If, in Equation 3, $t = t_{1/2}$, then n equals $n_0/2$, and it becomes clear that $t_{1/2}$ and k are related by the equation

$$\log_{10}1/2 = \frac{-kt_{1/2}}{2.3} \quad \text{or} \quad t_{1/2} = \frac{-0.301 \times 2.303}{-k} = \frac{0.693}{k} \tag{4}$$

If by some means we can find k for a given active isotope, we can find its half-life by Equation 4.

Devices have been developed that are sensitive to the high-energy particles emitted in the nuclear reactions that occur during radioactive decay processes. Upon entering such a device, usually called a counter, a high-energy particle causes an electric discharge in the counter. This discharge is automatically recorded on a digital readout meter. In a typical experiment the active sample is placed near the counter, and the number of particles emitted in the direction of the detector is determined. If the number of particles entering the counter is measured as a function of time, we can measure in a relative way the number of active atoms in the whole sample; for example, if after 10 minutes the number of counts per minute is only 80 per cent of the number at the beginning of the experiment, there are only 80 per cent as many active nuclei in the sample as there were at the start. This means that if the counting rate is called A, the activity of the sample, Equation 3 can be written in terms of A as well as n, in a relative way, to give

$$\log_{10}\frac{n}{n_0} = \log_{10}\frac{A}{A_0} = -\frac{kt}{2.3} \text{ or } \log A = \log A_0 - \frac{kt}{2.3} \tag{5}$$

To determine the decay constant k for a given active nucleus, we need merely to measure the activity of a sample containing that kind of nucleus as a function of time. Substituting into Equation 5 at two different times and counting rates will permit the elimination of $\log A_0$ and the evaluation of k. A more accurate method would be to measure the activity A of the sample at several times and to make a graph of $\log A$ as a function of time. Since $\log A$ varies linearly with time, the slope of the graph of $\log A$ versus t should be constant and equal to $-k/2.3$.

In this experiment we will use the latter procedure to measure the half-life of a radioactive isotope. Since most naturally occurring radioactive isotopes have very long half-lives, and thus could not be studied by this method, we will use synthetic radioactive isotopes prepared in this laboratory. Many nuclei, when bombarded with slow neutrons, absorb the neutrons and so undergo nuclear reactions. The nuclei so formed are typically radioactive but do not decay by neutron emission. Rather, they emit electrons of high energy and are thereby converted to nuclei of one higher atomic number. Naturally occurring iodine is typical in its behavior in this regard:

On bombardment with slow neutrons: $^{127}_{53}\text{I} + ^{1}_{0}\text{n} \rightarrow ^{128}_{53}\text{I}$

Decay reaction: $^{128}_{53}\text{I} \rightarrow ^{128}_{54}\text{Xe} + ^{0}_{-1}\text{e}$

The electrons, which are called β particles in such reactions, can be readily detected in a counter and so used to measure the decay rate. The isotopes we shall be using have half-lives of the order of a few hours, so that they can be conveniently studied in the course of a laboratory period. The radiation emitted is of relatively low energy and of short duration, and is not particularly hazardous. You should, however, use due caution and not get any sample into your mouth or let it remain for any length of time on your skin.

EXPERIMENTAL PROCEDURE
WEAR YOUR SAFETY GLASSES WHILE
PERFORMING THIS EXPERIMENT

Obtain a test tube containing an unknown radioactive sample from the stockroom.

Your instructor will demonstrate the use of the radioactivity counter in your laboratory. Do not make any adjustments of the counter settings unless you are specifically directed to do so.

Pour your sample into the planchet provided and smooth it out with the spatula so that it lies in a layer of uniform thickness. Place the planchet in the counter; turn the counter on and measure the number of counts recorded in a one-minute interval. Record the time at which you started the counter. Remove the planchet from the counter and place an empty planchet in the counter. Record the number of background counts obtained in a one-minute interval with the empty planchet.

Wait about 15 minutes and repeat the procedure, noting the number of counts per minute and the time you begin to count. If there is not too much congestion from other students wishing to use the counter, measure the background counting rate a second time.

Repeat the procedure at about 15-minute intervals until you have obtained six counting rates at six different times. During the course of the period you should also obtain about four measurements of the background radiation counting rate.

When you have finished the experiment, dispose of your sample as directed by your laboratory instructor.

DATA AND CALCULATIONS: Determination of the Half-life of a Radioactive Isotope

Unknown no. _____

Count obtained in one minute	Time when counter was turned on	Total elapsed time	Corrected counting rate° = activity, A	$\log_{10}$ corrected counting rate = $\log_{10} A$
_____	_____	_____	_____	_____
_____	_____	_____	_____	_____
_____	_____	_____	_____	_____
_____	_____	_____	_____	_____
_____	_____	_____	_____	_____
_____	_____	_____	_____	_____

Background count _____ _____ _____ _____

Average background _____ cpm obtained in one minute.

On the graph paper on the following page, make a graph of $\log_{10} A$ as a function of elapsed time.

Slope of line obtained in the graph $= \dfrac{\Delta \log A}{\Delta t} =$ _____ $=$ _____

Slope of the line $= -\dfrac{k}{2.303}$; $k =$ _____ min^{-1}

By Equation 4, $t_{1/2} = \dfrac{0.693}{k} =$ _____ $=$ _____ min

---

°Subtract the average background counting rate from the counts per minute obtained with the sample to obtain the corrected counting rate, or the activity, A, of the sample.

DETERMINATION OF THE HALF-LIFE OF
A RADIOACTIVE ISOTOPE (Data and Calculations)

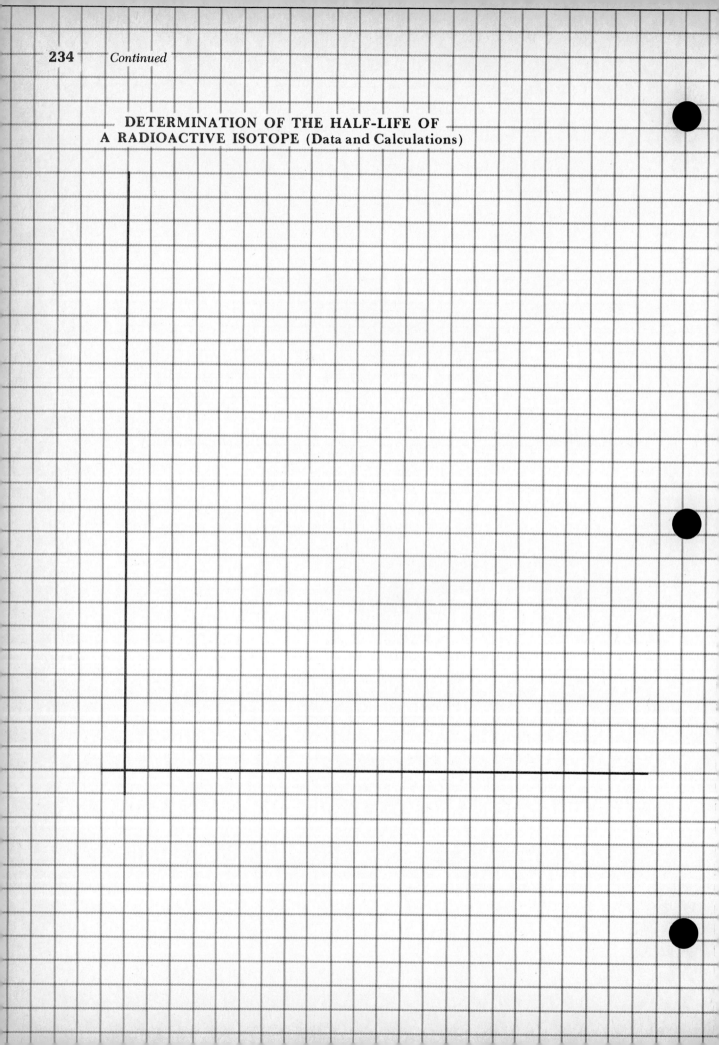

ADVANCE STUDY ASSIGNMENT: Determination of the Half-life of a
                            Radioactive Isotope

1.   In experiments of this sort the activity ratio A/A_0 can be taken to be equal to the active nuclei
ratio n/n_0. Why is this relation valid?

2.   A certain sample of a compound containing a radioactive isotope produced 820 counts in one
minute at the beginning of the experiment and a count of 530 in one minute when 57 minutes had
passed. If the background count was 20 cpm, what is the decay constant k and the half-life of the
radioactive species?

$k =$ _____ min^{-1}

$t_{1/2} =$ _____ min

29 • Preparation of Aspirin

One of the simpler organic reactions that one can carry out is the formation of an ester from an acid and an alcohol:

$$R-\overset{\overset{\text{O}}{\|}}{C}-OH \;+\; HO-R' \;\rightarrow\; R-\overset{\overset{\text{O}}{\|}}{C}-O-R' \;+\; H_2O \qquad (1)$$

an acid an alcohol an ester

In the equation, R and R′ are H atoms or organic fragments like CH_3, C_2H_5, or more complex aromatic groups. There are many esters, since there are many organic acids and alcohols, but they all can be formed, in principle at least, by Reaction 1. The driving force for the reaction is in general not very great, so that one ends up with an equilibrium mixture of ester, water, acid, and alcohol.

There are some esters which are solids because of their high molecular weight or other properties. Most of these esters are not soluble in water, so they can be separated from the mixture by crystallization. This experiment deals with an ester of this sort, the substance commonly called aspirin. Aspirin is the active component in headache pills and is one of the most effective, relatively nontoxic, pain killers.

Aspirin can be made by the reaction of the —OH group in the salicyclic acid molecule with the carboxyl (—COOH) group in acetic acid:

acetic acid salicylic acid aspirin

A better preparative method, which we will use in this experiment, employs acetic anhydride in the reaction instead of acetic acid. The anhydride can be considered to be the product of a reaction in which two acetic acid molecules combine, with the elimination of a molecule of water. The anhydride will react with the water produced in the esterification reaction and will tend to drive the reaction to the right. A catalyst, normally sulfuric or phosphoric acid, is also used to speed up the reaction.

acetic anhydride salicylic acid aspirin acetic acid

The aspirin you will prepare in this experiment is relatively impure and should certainly not be taken internally, even if the experiment gives you a bad headache.

There are several ways by which the purity of your aspirin can be estimated. Probably the simplest way is to measure its melting point. If the aspirin is pure, it will melt sharply at the literature value of the melting point. If it is impure, the melting point will be lower than the literature value by an amount that is roughly proportional to the amount of impurity present.

A more quantitative measure of the purity of your aspirin sample can be obtained by determining the per cent salicylic acid it contains. Salicylic acid is the most likely impurity in the sample since, unlike acetic acid, it is not very soluble in water. Salicylic acid forms a highly colored magenta complex with Fe(III). By measuring the absorption of light by a solution containing a known amount of aspirin in excess Fe^{3+} ion, one can easily determine the per cent salicylic acid present in the aspirin.

EXPERIMENTAL PROCEDURE

WEAR YOUR SAFETY GLASSES WHILE
PERFORMING THIS EXPERIMENT

Weigh a 50 ml Erlenmeyer flask on a triple beam or top loading balance and add 2.0 g of salicylic acid. Measure out 5.0 ml of acetic anhydride in your graduated cylinder, and pour it into the flask in such a way as to wash any crystals of salicylic acid on the walls down to the bottom. Add 5 drops of 85 per cent phosphoric acid to serve as a catalyst. *Both acetic anhydride and phosphoric acid are reactive chemicals which can give you a bad chemical burn, so use due caution in handling them.* If you get any of either on your hands or clothes, wash thoroughly with soap and water.

Clamp the flask in place in a beaker of water supported on a wire gauze on a ring stand. Heat the water with a Bunsen burner to about 75°C, stirring the liquid in the flask occasionally with a stirring rod. Maintain this temperature for about 15 minutes, by which time the reaction should be complete. *Cautiously*, add 2 ml of water to the flask to decompose any excess acetic anhydride. There will be some hot acetic acid vapor evolved as a result of the decomposition.

When the liquid has stopped giving off vapors, remove the flask from the water bath and add 20 ml of water. Let the flask cool for a few minutes in air, during which time crystals of aspirin should begin to form. Put the flask in an ice bath to hasten crystallization and increase the yield of product. If crystals are slow to appear, it may be helpful to scratch the inside of the flask with a stirring rod.

Collect the aspirin by filtering the cold liquid through a Buchner funnel using suction. Turn off the suction and pour about 5 ml of ice-cold distilled water over the crystals; after about 15 seconds turn on the suction to remove the wash liquid along with most of the impurities. Repeat the washing process with another 5 ml sample of ice-cold water. Draw air through the funnel for a few minutes to help dry the crystals and then transfer them to a piece of dry, weighed, filter paper. Weigh the sample on the paper to ± 0.1 g.

Test the solubility properties of the aspirin by taking samples of the solid the size of a pea on your spatula and putting them in separate 1 ml samples of each of the following solvents and stirring:

1. Toluene, $C_6H_5CH_3$, nonpolar aromatic
2. Hexane, C_6H_{14}, nonpolar aliphatic
3. Ethyl acetate, $C_2H_5OCOCH_3$, aliphatic ester
4. Ethyl alcohol, C_2H_5OH, polar aliphatic, hydrogen bonding
5. Acetone, CH_3COCH_3, polar aliphatic, nonhydrogen bonding
6. Water, highly polar, hydrogen bonding

To determine the melting point of the aspirin, add a small amount of your prepared sample to a melting point tube (made from 5-mm tubing), as directed by your instructor. Shake the solid down by tapping the tube on the bench top, using enough solid to give you a depth of about 5 mm. Set up the apparatus shown in Figure 29.1. Fasten the melting point tube to the thermometer with a small rubber band, which should be above the surface of the oil. The thermometer bulb and sample should be about 2 cm above the bottom of the tube. Heat the oil bath *gently*, especially

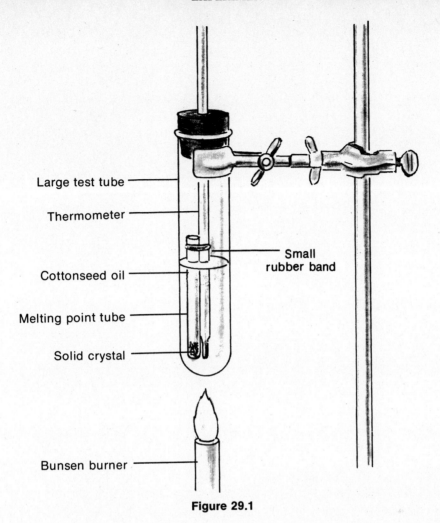

Large test tube

Thermometer

Small rubber band

Cottonseed oil

Melting point tube

Solid crystal

Bunsen burner

Figure 29.1

after the temperature gets above 100°C. As the melting point is approached, the crystals will begin to soften. Report the melting point as the temperature at which the last crystals disappear.

To analyze your aspirin for its salicylic acid impurity, weigh out 0.10 ± 0.01 g of your sample into a weighed 100-ml beaker. Dissolve the solid in 5 ml 95% ethanol. Add 5 ml 0.025 M $Fe(NO_3)_3$ in 0.5 M HCl and 40 ml distilled water. Make all volume measurements with a graduated cylinder. Stir the solution to mix all reagents.

Rinse out a spectrophotometer tube with a few ml of the solution and then fill the tube with that solution. Measure the absorbance of the solution at 525 nm. The absorbance measurement should be made *within 5 minutes* of the time the sample was dissolved in the ethanol, since aspirin will gradually decompose in solution, producing salicylic acid and acetic acid. From the calibration curve or equation provided, calculate the per cent salicylic acid in the aspirin sample.

Name _____ Section _____

DATA AND RESULTS: Preparation of Aspirin

Mass of salicylic acid used   _____g

Volume of acetic anhydride used   _____ml

Mass of acetic anhydride used
(density = 1.08 g/ml)

                                                                    _____g

Mass of aspirin obtained   _____g

Theoretical yield of aspirin

                                                                    _____g

Percentage yield of aspirin

                                                                    _____%

Melting point of aspirin   _____°C

Absorbance of aspirin solution   _____

Per cent salicylic acid impurity   _____%

Solubility properties of aspirin

    Toluene   _____   Ethyl alcohol_____

    Hexane   _____   Acetone   _____

    Ethyl acetate_____   Water   _____

S = soluble   I = insoluble   SS = slightly soluble

Comment on the likely ease of finding a good solvent for an organic solid with the general
structural complexity of aspirin.

 Name ______________ Section ______________

ADVANCE STUDY ASSIGNMENT: Preparation of Aspirin

1. Calculate the theoretical yield of aspirin to be obtained in this experiment, starting with 2.0 g of salicylic acid and 5.0 ml of acetic anhydride (density = 1.08 g/ml).

______________ g

2. If 1.9 g of aspirin were obtained in this experiment, what would be the percentage yield?

______________ %

3. The name acetic anhydride implies that the compound will react with water to form acetic acid. Write the equation for the reaction.

4. Identify R and R′ in Equation 1 when the ester, aspirin, is made from salicylic acid and acetic acid.

EXPERIMENT

30 • Preparation of a Synthetic Resin

In this experiment we will prepare and examine the properties of one of the most common polymers, polystyrene, made from styrene, $C_6H_5—CH{=}CH_2$. Styrene is relatively easy to polymerize; the plastic made from it is typically quite hard and transparent. In pure polystyrene, the chain is unbranched:

styrene   section of a polystyrene molecule

Polymers containing unbranched chains are thermoplastic, which means that they can be melted and then cast or extruded into various shapes. They also can usually be dissolved in some organic solvents, forming viscous liquids. If a polymer is cross-linked so that its chains are bonded together at regular or random positions, it will usually neither melt nor dissolve readily; such a material is called thermosetting and is usually polymerized in a mold in the shape of the article desired.

Styrene will polymerize to a crystalline solid if you simply heat it. The polymerization reaction itself evolves heat, however, and once the reaction gets started it tends to increase in rate and can get out of control; the simplest commercial process polymerizes styrene this way, and one of the important problems is to provide adequate cooling as the reaction proceeds.

We will polymerize styrene under somewhat different conditions, using an emulsion polymerization, in which the styrene is dispersed into droplets in water. In this process, temperature control is easy, and the polymer is produced in the form of easy-to-handle beads. By carrying out the reaction in the presence of divinylbenzene, which can react at two double-bonded positions, we will make a cross-linked polymer very similar in structure to that of an ion-exchange resin. Divinylbenzene is much like styrene except that there are two ethylene groups rather than one attached to each benzene ring. The compound has three isomers:

paradivinylbenzene   metadivinylbenzene   orthodivinylbenzene

The commercially available divinylbenzene which we use contains mainly the *para* and *meta* isomers, in about equal amounts.

245

The resin produced will have a structure similar to that indicated below:

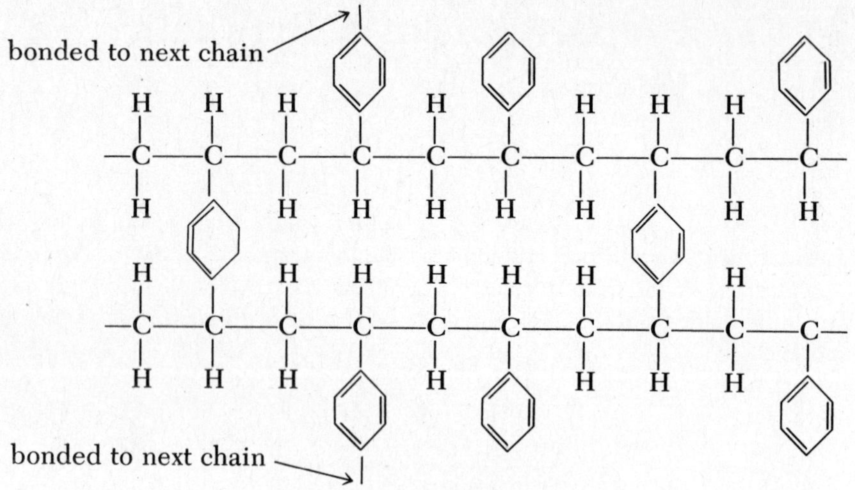

Relatively few divinylbenzene molecules are required for the cross-linking. The material produced is really a copolymer of styrene and divinylbenzene.

After preparing the polymer, we will compare its melting point and solubility properties with those of linear polystyrene.

EXPERIMENTAL PROCEDURE

WEAR YOUR SAFETY GLASSES WHILE PERFORMING THIS EXPERIMENT

Put 100 ml water in a 250 ml Erlenmeyer flask and heat the flask in a water bath set up as shown in Figure 30.1. When the water in the flask is at about 60°C, slowly add 1.0 g of starch, with stirring; continue stirring and heating until the solution is uniform and the starch completely dispersed.

While the water is heating, measure out 10 ml of styrene and 1 ml of divinylbenzene into a small beaker. Your instructor will add 100 mg benzoyl peroxide (very reactive!) to the beaker. Stir to dissolve the solid and initiate the reaction; keep the beaker in ice water until the starch solution has been prepared. When the starch solution is ready and at about 80°C, remove it from the water bath and slowly, with swirling, add the styrene solution; stopper the flask with a cork, not *too* tightly, to minimize vaporization of the styrene. Continue to swirl the liquid for about 20 seconds to disperse the styrene as small droplets in the starch. Do not shake the flask, since we do not wish to produce a true emulsion, just a dispersion of droplets. Put the flask back in the water bath and heat the mixture in the 80 to 90°C water for about an hour, during which time it should polymerize completely. Stir every few minutes with your stirring rod to keep the droplets dispersed. If all goes well, the polymer will form as small beads, varying in size from very small up to about 2 mm in diameter.

While the polymerization is proceeding, tear some polystyrene foam from a coffee cup into small pieces and use it to fill 3 small test tubes. The polymer in the foam is essentially linear polystyrene with few branches and no cross-links. Put 2 ml toluene in one of the test tubes and 2 ml acetone in the second; shake to get the foam wet with solvent.

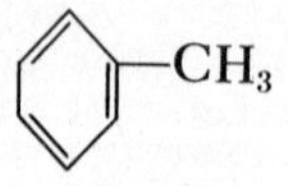

toluene

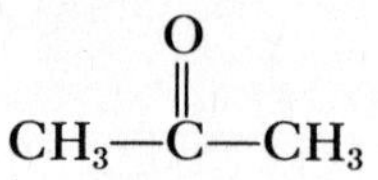

acetone

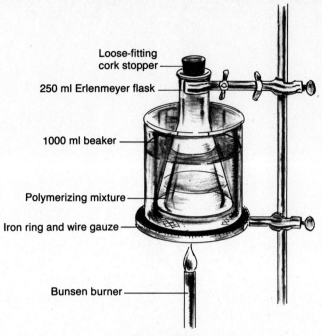

Loose-fitting
cork stopper

250 ml Erlenmeyer flask

1000 ml beaker

Polymerizing mixture

Iron ring and wire gauze

Bunsen burner

Figure 30.1

Heat the third test tube gently in the Bunsen flame, noting whether the polymer melts or decomposes. Estimate the temperature at which the change occurs, but don't try to measure it. Poke the material with your stirring rod to aid in establishing its viscosity. Shake the tubes containing solvent and note whether the polymer has dissolved; if it hasn't, put the tubes in the hot water in the bath to speed up the solution process. Do not boil the solvent, however. Note the properties of the final mixture, particularly viscosity and clarity of solution.

If you are able to obtain a solution with either solvent, pour a drop of the solution on to some water in a 600 ml beaker. Blow gently on the surface to evaporate the solvent completely. Pick up the film with a stirring rod and note its thickness and strength.

When the polymerization reaction is finished, remove the flask from the water bath and pour the slurry into a 600 ml beaker half full of distilled water. Stir and then let the beads settle. Decant the liquid and wash twice more to remove any residual starch. Pour the beads out on a paper towel and, when they are dry, weigh them.

Put a few of the beads in small test tubes and test them as before for solubility in toluene and acetone. Try to melt the beads; compare their behavior on heating with that of the polystyrene foam.

CAUTION: In this experiment the laboratory should be well ventilated, since we use several volatile organic liquids. If it is convenient, work in a hood or open the windows in the lab. Avoid contact with the liquids and their vapors. If you keep the polymerizing mixture moderately well stoppered, enough to let just a bit of air in or out, there should be very little styrene vapor in the air. Open the flask momentarily while you stir the mixture.

Name _____ Section _____

DATA AND OBSERVATIONS: Preparation of a Synthetic Resin

Mass of styrene (density = 0.90 g/ml)   _____g

Mass of divinylbenzene (density = 0.90 g/ml)   _____g

Mass of resin   _____g

Theoretical yield   _____g

Percentage yield   _____%

Properties of polystyrene and prepared resin

	Polystyrene (foam cup)	Prepared resin
Behavior on heating	_____	_____
	_____	_____
Solubility in toluene	_____	_____
Solubility in acetone	_____	_____

How do you explain the difference in solubility of polystyrene in the two solvents?

Comment on the effects of cross-linking on the properties of polystyrene.

ADVANCE STUDY ASSIGNMENT: Preparation of a Synthetic Resin

1.   Polypropylene is made by addition polymerization of propylene, $CH_2\!\!=\!\!C\!\!-\!\!H$. Sketch a section of the polypropylene molecule.
$$\overset{\displaystyle |}{\underset{\displaystyle CH_3}{}}$$

2.   How much polypropylene could theoretically be made from 100 g of propylene?

_____ g

3.   What per cent by mass of polypropylene is carbon?

_____ %

4.   Polystyrene foam such as that used in coffee cups is made by a procedure very analogous to that used in this experiment. Can you suggest how the foamable polymer might be made and how it would be converted to the form of a foam coffee cup?

QUALITATIVE ANALYSIS

Introduction to the Qualitative Analysis of Cations and Anions

The remaining experiments in this manual are devoted to that area of chemistry called qualitative analysis, in which one studies the methods by which one can determine the nature, but not the amount, of the species in a mixture. In these experiments you will usually be asked to analyze unknown solutions containing, for the most part, metallic cations and inorganic anions in aqueous solutions. The scheme of analysis we will use will allow for the identification of 22 cations and 15 anions.

Since the general problem of analysis of a complex mixture of ions is by no means simple, the scheme of analysis is broken down into several parts; each part involves a fairly small group of ions that can be isolated from the general mixture, or at least studied as a separate set, on the basis of some property common to the ions in the group.

Analysis of Cations. The procedure for determining which cations are in a solution is somewhat more systematic than that used with anions and so is dealt with first. In our scheme we separate the cations into four Groups. By means of selective precipitations, these Groups are removed one at a time and in a definite order from the general mixture. The cations in Group I are removed first, leaving the cations in Groups II, III, and IV in solution; from that solution the Group II cations are then precipitated, leaving Groups III and IV in solution, and so on.

The members of each Group, and the conditions under which that Group is separated, are as follows:

Group I	Ag^+, Hg_2^{2+}, Pb^{2+}	Precipitated as chlorides under strongly acidic conditions
Group II	Cu^{2+}, Cd^{2+}, Bi^{3+}, Sn^{2+} and Sn^{4+}, Hg^{2+}, Sb^{3+} and Sb^{5+}, (Pb^{2+})	Precipitated as sulfides under mildly acidic conditions
Group III	Al^{3+}, Zn^{2+}, Cr^{3+}, Fe^{2+} and Fe^{3+}, Ni^{2+}, Co^{2+}, Mn^{2+}	Precipitated as sulfides or hydroxides under slightly basic conditions
Group IV	Ba^{2+}, Ca^{2+}, Mg^{2+}, Na^+, K^+, NH_4^+	Remain in solution after precipitation of Group III cations

After separation, the cations within a Group are further resolved by means of a series of chemical reactions into sets of soluble and insoluble fractions, until the resolution is sufficient to allow identification of each cation by one or more tests specific to that ion. During the course of these separations, all the different kinds of chemical reactions we have studied in this course will be used; in addition to precipitations, we will employ acid-base reactions, complex ion formation, oxidation-reduction, and combinations of these kinds of reactions to accomplish the analyses.

Analysis of Anions. In carrying out the qualitative analysis of anions we proceed in a manner somewhat similar to that outlined above. The anions fall into four Groups, each with a characteristic property common to the anions in the Group. Here, however, the anions in a Group are either separated from the general mixture before analysis or dealt with by "spot tests" designed to detect the ion directly in the general mixture. A sequence of Group precipitations such as is used in cation analysis is not easily accomplished with the anions, making anion analysis somewhat less systematic. It is also made more difficult by the fact that some anions tend under certain conditions to react with other anions and thus to be converted to other species.

The members of the anion Groups, along with their characteristic properties, are listed below:

Silver Group	Cl^-, Br^-, I^-, SCN^-	Precipitated by Ag^+ ion under strongly acidic conditions
Calcium-Barium-Iron Group	SO_4^{2-}, PO_4^{3-}, CrO_4^{2-}, $C_2O_4^{2-}$	Precipitated by Ca^{2+} ion or Ba^{2+} ion at proper pH
Soluble Group	NO_3^-, NO_2^-, ClO_3^-, $C_2H_3O_2^-$	Not easily precipitated
Volatile Acid Group	CO_3^{2-}, SO_3^{2-}, S^{2-}	Evolve gases on acidification

For the most part we will limit each qualitative analysis experiment to the study of the ions in one or, at most, two groups. At the beginning of each experiment there is a discussion of the properties of each of the ions in the Group, including the properties actually used in the analysis scheme, as well as others. This is followed by a detailed procedure by which a solution containing the ions in the Group can be analyzed for the presence of those ions. The laboratory assignments for each experiment include carrying out the procedure with a known solution containing all of the ions in the Group and then using that procedure to analyze an unknown solution that may contain any of those ions. Following this you may be asked to devise your own scheme for the analysis of a sample of limited possible composition, containing only a few ions; for this problem you may use a simplified version of the standard procedure or a scheme based on other properties of the ions included in the discussion section.

In the experiments involving single groups of ions, the solutions used will not contain any interfering ions of opposite charge. The sources of cations in solution will be nitrates or chlorides, and the anion systems will be obtained from sodium or potassium salts. In general mixtures, where the cations and anions may react with each other and there is some likelihood of anion-anion reactions, the situation is considerably more complicated. In the last experiment we will consider mixtures of this sort and will ask you to do an analysis for both cations and anions. In that experiment we will work with solid mixtures and will describe methods for preparing solutions prior to analysis as well as for avoiding difficulties caused by unwanted ion-ion reactions.

In addition to giving you the nontrivial (and perhaps sometimes frustrating) experience of following a rather complicated set of directions for carrying out series of chemical reactions, studies in qualitative analysis offer you an opportunity for working with many different kinds of chemical substances and learning a good deal about their behavior under varying conditions. The procedures used illustrate very well how the principles of chemical equilibrium can be applied to make systems react in desired ways to furnish needed information. If you examine the procedures carefully, you will find that they are fairly easily understood in terms of the laws of chemistry and the properties of the ions being studied, and that, given time, you could have developed them yourself. Such an examination should make it clearer to you what is accomplished in each stage in the procedure. If you know why you are asked to wash a precipitate a given way, or why a mixture must be heated or boiled, you are more likely to perform the operation properly.

LABORATORY PROCEDURES

Good laboratory technique in qualitative analysis is important if you are to get results like those described in the procedures. Sloppy, careless work will produce precipitates of the wrong color, or precipitates where you should get solutions, and in general will make your analytical life less pleasant and less successful than it might be. Experience is certainly crucial to good technique, but a little advice before you begin these experiments might be useful in helping you get started properly.

Our qualitative analyses will be carried out on what is commonly called the semimicro level. The amounts of chemicals we use are neither large, of the order of grams, nor very small, of the order of micrograms. The sample solutions are typically about 0.1 M, and we work with volumes of about 1 ml, which means that there will be, on the average, about 10 mg of solute in the sample. The identification tests for the most part will be ineffective if you have less than 1 or 2 mg of solute present, so you must carry out the procedures carefully enough to avoid losing the major portion of any component somewhere during the analysis.

Separation of Solid Precipitates from Solutions.   One of the reasons that we work at the semimicro level is that at that level one can very conveniently use small centrifuges to separate a precipitate from a solution. The common laboratory centrifuge will accept a small semimicro test tube holding about 8 ml; such tubes (13 mm $\times$ 100 mm) are large enough to hold the reaction mixtures that we will prepare, and they will be used almost exclusively in our work.

In a previous experiment you may have used centrifugation to remove solid PbI_2 from solution. In general, to centrifuge a sample, put the test tube containing the precipitate and solution into one of the locations in the centrifuge and another test tube containing a similar amount of water on the opposite side of the rotor. (You may find that such a tube is permanently taped in the centrifuge, in which case you should put your test tube in the opposite slot.) Turn on the centrifuge and let it run for about 30 seconds, during which time most precipitates will settle to a compact mass in the bottom of the tube. If your sample is still suspended, centrifuge it again; if this still does not work, you may find it helpful to heat the sample in the water bath for a few minutes, thereby promoting formation of larger crystals of solid, which tend to centrifuge out more easily.

Measuring Out Amounts of Solutions or Solids.   A typical step in a procedure is to add 1 ml of a reagent, or 0.1 g of a solid, to a solution. Perhaps surprisingly, this is almost *never* done by using a graduated cylinder or a balance to measure out the amount of liquid or solid. The direction given means to add about 1 ml, say ±0.2 ml, or about 0.1 g $\pm$ 0.02 g. Measuring out an approximate volume of this sort is easy, and fast, if you know roughly the volume that 1 ml occupies in a semimicro test tube. This can be determined by putting about 5 ml of water in a small graduated cylinder and then pouring the water into the test tube, 1 ml at a time. After you have done this a few times you should be able to judge the increase in level that corresponds to 1 ml, and, indeed, to 0.5 ml. In general, use the medicine dropper in the reagent bottle to transfer solution directly to the test tube containing your sample. *Never* contaminate a reagent solution by dipping your own eye dropper or pipet into it. Add distilled water directly from your squeeze bottle, again estimating volume from the change in level of the liquid. If you wish, you can measure volume by counting drops from the medicine dropper. Most medicine droppers deliver about the same volume per drop, and you can check your own dropper to see how many drops equal 1 ml. Typical droppers deliver 10 to 15 drops per milliliter, with 12 drops being a good average, at least in our laboratories.

Dispensing a given approximate amount of solid can be accomplished in a way similar to that used with liquids. One usually adds a small amount of solid from the tip of a small spatula, and the problem is to determine the space that, say, 0.1 g of solid occupies. To find out, work with a sample of a typical solid, such as powdered limestone, $CaCO_3$, and a sensitive top-loading or triple-beam balance. Put a small beaker on the balance and weigh it. Then add portions of the solid on the end of the spatula, weighing each until you can add a sample that you are sure weighs 0.1 g $\pm$ 0.02 g. If you assume that an equal volume of any other solid will weigh about 0.1 g, you will not go too far wrong.

Precipitating a Solid.   One of the most common reactions in qualitative analysis is the precipitation of substances from solution. To accomplish a precipitation, add the indicated amount of precipitating reagent to the sample solution and stir well with your glass stirring rod. Heat the solution in the water bath if so directed. Since some precipitates form slowly, they must be given time to form completely. It is hard to overstir, but students frequently do not mix reagents thoroughly enough. When the precipitation is believed to be complete, centrifuge out the solid, and before decanting the liquid, add a drop or two of the precipitating reagent, just to make sure that you added enough precipitant the first time. The directions usually specify enough reagent to furnish an excess, but it is good insurance to test, just the same.

Washing a Solid.   The liquid decanted from the test tube after centrifuging out a solid does not in general contain any of the solid and may be used directly in a following step. The solid remaining in the tube, however, has residual liquid around it. Since this liquid contains ions that may interfere with further tests on the solid, they must be removed. This is accomplished by diluting the liquid with a wash liquid, often water, which does not interfere with the analysis, dissolve the solid, or precipitate any substances from solution.

To wash the solid, add the indicated amount of wash liquid to the test tube and mix well with your glass stirring rod, dispersing the solid well in the wash liquid. Simply pouring in the wash liquid and pouring it out again is not effective. After mixing thoroughly, centrifuge out the solid and decant the wash liquid, which may usually be discarded. The washing operation, in key separations, is best done twice, because an uncontaminated precipitate tends to give much better results than one mixed with even a little chemical trash.

Heating a Mixture.   Many reactions are best carried out when the reagents are hot. We *never*, or almost never, heat a test tube containing a reaction mixture directly in a Bunsen flame. It is more convenient and much more considerate of your neighbor to heat the test tube in a water bath. A 250-ml beaker containing about 150 ml of water makes a very adequate water bath.

When you are following a procedure in which you need to heat a solution, keep the water bath hot or simmering by using a small flame to heat the beaker on a piece of asbestos-covered wire gauze. Then the bath is ready whenever required and you do not have to wait 10 minutes for the water to heat. The test tube, or tubes, can be put directly into the bath, supported against the wall of the beaker. In a water bath you can bring the mixture under study to approximately 100°C without boiling it. If you heat the sample in an open flame, it inevitably bumps out of the tube and onto the laboratory bench, or onto you or someone else. Therefore, *do not* heat test tubes containing liquids over an open flame.

Evaporating a Liquid.   In some cases it will be necessary to boil down a liquid to a small volume. This may be done to concentrate a species or to remove a volatile reagent, or perhaps because the reaction will proceed readily only in a boiling solution. When this operation is necessary, we perform the boiling in a small 50-ml beaker on a square of asbestos-covered wire gauze and use a small Bunsen flame judiciously applied to maintain controlled, gentle boiling. Since the volumes involved usually are of the order of 3 or 4 ml, overheating can easily occur.

If you are supposed to stop the evaporation at a volume of 1 ml or so, make sure that you do not heat to dryness, because you may decompose the sample or render it inert. Since concentrated solutions or slurries tend to bump, it is often helpful to encourage smooth boiling by scratching the bottom of the beaker with a stirring rod as the boiling proceeds. Good judgment in boiling down a sample is important, so use it.

Occasionally the liquid that is evaporated is highly acidic with HCl or HNO_3, and the boiling causes evolution of noticeable amounts of toxic gases into the laboratory. In some cases the amounts are small enough to be ignored. If you or anyone else notices that bothersome vapors are escaping from your boiling mixture, transfer your operation to the hood. In some laboratories each station has its own small hood; in such a case, carry out all your evaporations under your hood.

Transferring a Solid.   Sometimes it is necessary to transfer a solid from the beaker in which it was prepared to a test tube for centrifuging, or from the test tube to the beaker. The amount of solid involved is never large, of the order of 50 mg at most. We always perform such transfers in the presence of a liquid, which serves as a carrier.

When you are ready to make the transfer, stir the solid well into the liquid, forming a slurry, and then, without delay, pour the slurry into the other container. The transfer is not quantitative, but if done properly, you can move 90 per cent of the sample to the other container. Forget about the rest. In general, we do not attempt to transfer wet solids with a spatula, because it is not easy, and all too often the spatula reacts with the liquid present, contaminating it.

Handling a Stirring Rod.   You will find that a glass stirring rod is a very useful tool and that you need it in nearly every step in each procedure. The problem is that each time it is used, it gets wet with the solution being treated, and then it must be cleaned before it can be used again. This is a minor problem, but a bothersome one. The best solution we have found is to keep a 400-ml beaker full of distilled water handy as a storage place for stirring rods. After you have used a rod, swirl it around in the water in the beaker to clean it, and then leave it in the water. Solutes and solids accumulate, but they really amount to tiny traces when diluted in the water. Change the water occasionally to ensure that no significant contamination occurs. We have found this procedure for handling stirring rods to work very well and have had no spurious test results yet.

Adjusting the pH of a Solution.   One of the most important variables controlling chemical reactions is the pH of the solution. Frequently it is necessary to make a basic solution acidic, or vice versa, in order to make a desired reaction take place. For instance, if you are directed to add 6 M HCl to an alkaline mixture until it is acidic, you should proceed in the following way. Knowing how the alkaline solution was prepared, make a quick mental calculation of about how much acid is needed—1 drop, 1 ml, or perhaps more. Then add the acid, drop by drop, until you think the pH is about right. Mix well with your stirring rod and then touch the end of the rod to a piece of blue litmus paper on a piece of paper towel or filter paper. If the color does not change, add another drop or two of acid, mix and test again. Frequently the system changes its character at the neutral point; a precipitate may dissolve or form, or the color may change. In any event, add enough acid so that, after mixing, the litmus paper turns red when touched with the stirring rod. Similarly, if you are told to make a solution basic with 6 M NH_3 or 6 M NaOH, add the reagent, drop by drop, until the solution, after being well mixed, turns red litmus blue. Adjustment of pH is not difficult, but it must be done properly if the desired reaction is to occur.

EXPERIMENT

31 • Qualitative Analysis of the Group I Cations: Ag^+, Pb^{2+}, Hg_2^{2+}

SEPARATION OF THE GROUP I CATIONS

In the classical qualitative analysis scheme, the first ions which are determined are Ag^+, Pb^{2+}, and Hg_2^{2+}. Of all the common cations, only these form insoluble chlorides under acidic conditions. These cations comprise Group I in the scheme and are separated from a general salt solution by precipitation of their chlorides at a pH of about zero. The chloride precipitate is then analyzed for the possible presence of silver, lead, and mercury(I) on the basis of the characteristic properties of those cations.

PROPERTIES OF THE GROUP I CATIONS

Ag^+. Silver has only a few water soluble salts, of which the nitrate is certainly the most common. Most of the insoluble silver salts dissolve in cold 6 M HNO_3, the main exceptions being the silver halides, AgSCN, and Ag_2S. Silver ion forms many stable complexes; of these, the best known is probably the $Ag(NH_3)_2^+$ ion. This complex is sufficiently stable to be produced when AgCl or AgSCN is treated with 6 M NH_3; the reaction which occurs is useful for dissolving those solids. AgBr and AgI are less soluble than AgCl; AgBr will go into solution in 15 M NH_3, but AgI is so insoluble that it will not. The silver thiosulfate complex ion, $Ag(S_2O_3)_2^{3-}$, is extremely stable, and is important in photography, where it is formed in the "fixing" reaction in which AgBr is removed from the developed negative.

Pb^{2+}. Lead nitrate and acetate are the only well known soluble lead salts. Lead chloride is not nearly as insoluble in water as are the chlorides of silver and mercury(I), and becomes moderately soluble if the water is heated. $PbSO_4$ is one of the relatively few insoluble sulfates. Lead forms a stable hydroxide complex ion and a weak chloride complex. Although lead ordinarily has an oxidation number of $+2$, there are some Pb(IV) compounds, of which the most common is PbO_2 (brown); this compound is insoluble in most reagents, but will dissolve in 6 M HNO_3 to which some H_2O_2 has been added.

Hg_2^{2+}. Mercury(I) has only one soluble salt, the nitrate, and even with this compound, excess HNO_3 must be present to keep basic Hg(I) salts from precipitating. The Hg(I) ion, sometimes called mercurous ion, is relatively unstable; it will slowly oxidize to Hg(II) if exposed to air, and can be reduced to the metal by reducing cations (e.g., Sn^{2+}). In the presence of species that form Hg(II) complex ions or insoluble salts, mercurous ion often undergoes a disproportionation reaction to Hg (solid, black) and the Hg(II) compound. Mercury(I) chloride has a very characteristic reaction with ammonia:

$$Hg_2Cl_2(s) + 2\,NH_3(aq) \rightarrow Hg(s) + HgNH_2Cl(s) + NH_4^+(aq) + Cl^-(aq)$$

In Table 31.1 we have summarized some of the general solubility properties of the Group I cations. There is a good deal of information in the table, and you should learn to properly interpret and use the information it contains. For example, in the entry for Ag^+ and CrO_4^{2-} we have C, A (dk red). This means that if water solutions containing Ag^+ and CrO_4^{2-} ions are mixed, we obtain a precipitate of dark red Ag_2CrO_4; this substance is not soluble in water but would dissolve in an acidic solution such as 6 M HNO_3 (A), which would not contain a precipitating anion, and in

259

TABLE 31.1 SOLUBILITY PROPERTIES OF THE GROUP I CATIONS

	Ag^+	Pb^{2+}	Hg_2^{2+}
Cl^-	C, A^+ (white)	HW, C, A^+ (white)	O^+ (white)
OH^-	C, A (brown)	C, A (white)	D (black)
SO_4^{2-}	S^-, C (white)	C (white)	S^-, A (white)
CrO_4^{2-}	C, A (dk red)	C (yellow)	A (orange)
CO_3^{2-}, PO_4^{3-}	C, A (white)	C, A (white)	A (white)
S^{2-}	O (black)	O (black)	D (black)
Complexes	NH_3, $S_2O_3^{2-}$	OH^-	—

Key:
S   soluble in water, >0.1 mole/liter
S^- slightly soluble in water, ~0.01 mole/liter
HW   soluble in hot water
A   soluble in acid (6 M HCl or other non-
     precipitating, nonoxidizing acid)
I   insoluble in any common solvent

A^+ soluble in 12 M HCl
B   soluble in hot 6 M NaOH containing S^{2-} ion
O   soluble in hot 6 M HNO_3
O^+ soluble in hot aqua regia
C   soluble in solution containing a good
     complexing ligand
D   unstable, decomposes

solutions containing ligands which form stable complexes with silver (C). The entry under Complexes tells us that those ligands normally encountered in qualitative analysis include NH_3 and $S_2O_3^{2-}$.

GENERAL SCHEME OF ANALYSIS

The procedure used for the qualitative analysis of the Group I cations makes simple, straightforward use of some of the properties of these ions. Following separation of the chloride precipitate, the solid is treated with hot water to dissolve any $PbCl_2$ that is present. The hot solution containing Pb^{2+} ion is then mixed with a solution of K_2CrO_4; if lead is present, a yellow precipitate of $PbCrO_4$ will be produced. The remaining chloride precipitate is treated with 6 M NH_3. If Hg(I) is present, a black precipitate containing Hg(s) will form. Any silver chloride present will dissolve as $Ag(NH_3)_2^+$; after separation from the solid, the solution of the silver complex ion is acidified, which destroys the complex and reprecipitates white AgCl. In the presence of large amounts of Hg_2^{2+}, silver ion may be reduced to metallic silver on addition of NH_3 to the chloride precipitate; under such conditions the black precipitate which is obtained in that step is dissolved in aqua regia, and the resulting solution is tested for silver.

PROCEDURE FOR ANALYSIS OF GROUP I CATIONS

WEAR YOUR SAFETY GLASSES WHILE
PERFORMING THIS EXPERIMENT

Unless told otherwise, you may assume that 10 ml of your sample contains the equivalent of about 1 ml of 0.1 M solutions of the nitrate salts of one or more of the Group I cations, plus possibly ions from Groups II, III, and IV. Roughly speaking, this amounts to about 10 mg of each cation present. This is a sufficient amount for good qualitative tests, *as long as you do not lose any component cations* by improperly carrying out or interpreting any step.

Step 1. To 3 ml of your sample in a test tube, add 0.5 ml 6 M HCl. Stir well and centrifuge. Decant the liquid, which may contain ions from groups to be discussed later, into a test tube and save it, if necessary, for further analysis; to make sure precipitation of Group I cations was complete, add 1 drop of 6 M HCl to the liquid. Wash the precipitate with 2 ml water and 3 drops of 6 M HCl. Stir well. Centrifuge and discard the wash liquid. Wash the precipitate again with water and HCl; centrifuge and discard the wash.

Step 2. To the precipitate from Step 1, which contains the chlorides of the Group I cations, add about 4 ml water. Heat in the boiling water bath for at least three minutes, stirring constantly. Centrifuge quickly and decant the liquid, which may contain Pb^{2+}, into a test tube.

Step 3. *Confirmation of the presence of lead.* To the liquid from Step 2 add 2 drops of 6 M acetic acid and 3 or 4 drops of 1 M K_2CrO_4. The formation of a yellow precipitate of $PbCrO_4$ confirms the presence of lead. Centrifuging out the solid may help with the identification, because the liquid phase is orange.

Step 4. *Confirmation of the presence of mercury.* If lead is present, wash the precipitate from Step 2 with 6 ml water in the boiling water bath. Centrifuge and test the liquid for Pb^{2+}. Continue the washings until no positive reaction to the lead test is obtained. To the washed precipitate add 2 ml 6 M NH_3 and stir well. A black or dark gray precipitate establishes the presence of the mercury(I) ion. Centrifuge and decant the liquid, which may contain $Ag(NH_3)_2^+$, into a test tube.

Step 5. *Confirmation of the presence of silver.* To the liquid from Step 4 add 3 ml 6 M HNO_3. Check with litmus to see that the solution is acidic. A white precipitate of AgCl confirms the presence of silver.

Step 6. *Alternative confirmation of the presence of silver.* If the test for silver ion was inconclusive, and mercury was present, add 1 ml 12 M HCl and 0.5 ml 6 M HNO_3 to the precipitate from Step 4. Heat in the water bath until solution is essentially complete. Pour the liquid into a 50-ml beaker and boil it gently for a minute. Add 5 ml water. If a white precipitate forms, it is probably AgCl. Centrifuge and discard the liquid. To the precipitate, add 0.5 ml 6 M NH_3; with stirring, the precipitate should dissolve. Add 1 ml 6 M HNO_3; if silver is present, a precipitate of AgCl forms.

COMMENTS ON PROCEDURE FOR ANALYSIS OF GROUP I CATIONS

Step 1. In this step the cations of Group I are precipitated as their chlorides. Both silver and lead ions can form chloride complexes in solutions when the chloride ion concentrations are high. In our procedure, $[Cl^-]$ is about 1 M, which decreases the salt solubilities by the common ion effect, but is not great enough to cause appreciable amounts of the complex ions to form. If, of the Group I cations, only lead is present, it may not precipitate unless its concentration is ~ 0.1 M or greater.

Step 2. The $PbCl_2$ is sufficiently soluble in hot water to allow its separation from the other chlorides by simply heating the precipitate, with mixing, in water. The centrifuging should be done quickly to avoid reprecipitation of $PbCl_2$ on cooling.

Step 3. We acidify the liquid to minimize precipitation of other chromates from residual amounts of ions in other groups. The liquid is orange because of the conversion of CrO_4^{2-} to $Cr_2O_7^{2-}$ in the acid medium.

Step 4. If lead chloride is not completely removed, it is converted to a white, basic, insoluble salt on addition of NH_3. This could cause confusion, but should not interfere with later identifications. The AgCl dissolves readily in 6 M NH_3, with formation of the silver ammonia complex ion. If Hg_2Cl_2 is present, it reacts with NH_3, forming black Hg and white insoluble $HgNH_2Cl$. The mixture is black or dark gray if mercury(I) ion is present.

Step 5. The silver ammonia complex ion is destroyed by acid, and the released silver ion precipitates with the chloride ion in solution. The formation of white AgCl is definitive evidence for the presence of silver.

Step 6. If both mercury(I) ion and silver ion are present, the $Ag(NH_3)_2^+$ and mercury, both present in Step 4, tend to undergo an oxidation-reduction reaction:

$$2 \ Ag(NH_3)_2^+(aq) \ + \ Hg(s) \ + \ Cl^-(aq) \rightarrow 2 \ Ag(s) \ + \ HgNH_2Cl(s) \ + \ NH_4^+(aq) \ + \ 2 \ NH_3(aq)$$

If sufficient mercury is present, nearly all the silver may be reduced and a very doubtful test for Ag^+ obtained in Step 5. If it appears that this is the case, carrying out Step 6 should be helpful. On being dissolved in aqua regia, the mercury exists as $HgCl_4^{2-}$ and the silver as $AgCl_2^-$. Adding water to the strongly acidic solution reprecipitates white AgCl, which can then be separated from the $HgCl_4^{2-}$ by centrifuging. The solution of AgCl in NH_3 and reprecipitation on addition of HNO_3 is definitive confirmation of the presence of silver.

LABORATORY ASSIGNMENTS

Perform one or more of the following, as directed by your instructor:

1. Make up a sample containing about 1 ml of 0.1 M solutions of the nitrate salts of each of the cations in Group I. Go through the standard procedure for analysis of the Group I cations, comparing your observations with those that are given. Then obtain a Group I unknown from your instructor and analyze it to determine possible presence of Ag^+, Pb^{2+}, and Hg_2^{2+}. On a Group I flow chart, indicate your observations and conclusions about the unknown and submit the completed chart to your instructor.

2. If a sample may contain *only* cations from Group I, it is possible to analyze it by several procedures that are quite different from the one given in this book. Develop a scheme for analysis of such a sample, starting with the addition of 6 M NaOH in excess. Draw a complete flow chart for your procedure, indicating reagents to be added at each step and the formulas and colors of all species present during the course of the analysis. Test your procedure with a Group I known, and then use your method to analyze an unknown sample. In another color, indicate on your flow chart your observations on the unknown and your conclusions regarding its composition.

Note: In this and all succeeding laboratory assignments in this manual, the first assigned problem involves your acquiring some familiarity with the standard procedure for analysis. Succeeding problems require your developing and using your own schemes for analyzing particular unknown mixtures. In setting up your procedures, you may use steps from the standard schemes, but you should also examine the characteristic properties of the individual ions to see if some of them might be profitably used in your scheme. In all probability, the best method for analyzing any given limited mixture of ions will be partly based on the standard procedure and partly based on ion properties that were not made use of in the standard approach.

Outline of Procedure for Analysis of Group I Cations (Group I Flow Chart)

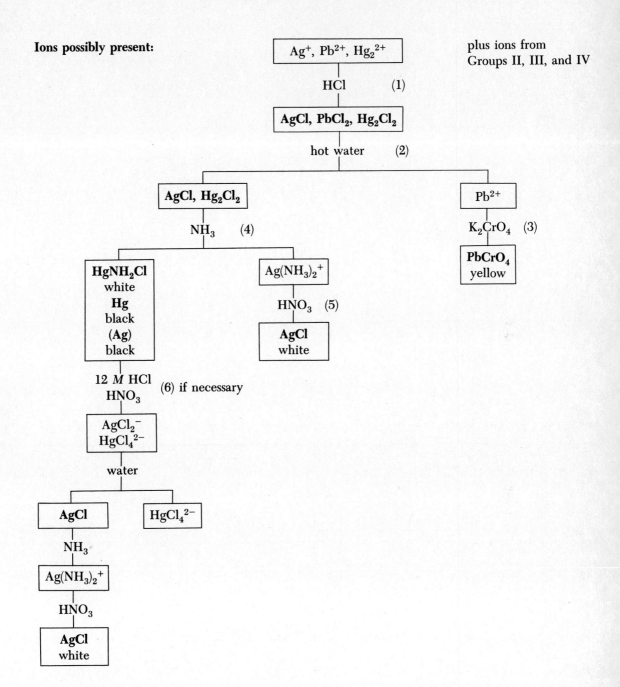

Ions possibly present:

Ag⁺, Pb²⁺, Hg₂²⁺

plus ions from Groups II, III, and IV

OBSERVATIONS: **Analysis of Group I Cations**

Assignment 1   Flow Chart Showing Behavior of Unknown

Assignment 2   Flow Chart

ADVANCE STUDY ASSIGNMENT: Analysis of Group I Cations

1. Write balanced net ionic equations for the following reactions:
    a. The precipitation of the chloride of Hg_2^{2+}.

    b. The confirmatory test for Pb^{2+}.

    c. The dissolving of AgCl in aqueous ammonia.

2. You are given an unknown solution that contains only one of the Group I cations and no other metallic cations. Develop the simplest procedure you can think of to determine which cation is present. Draw a flow chart showing the procedure and the observations to be expected at each step with each of the possible cations.

3. A solution may contain Ag^+, Pb^{2+}, and Hg_2^{2+}. A white precipitate forms on addition of 6 M HCl. The precipitate is partially soluble in hot water. The residue turns black on addition of ammonia. Which of the ions are present, which are absent, and which remain undetermined? State your reasoning.

                                            Present _____

                                            Absent   _____

                                            In doubt _____

EXPERIMENT

32 • Qualitative Analysis of the Group II Cations: Cu^{2+}, Bi^{3+}, Hg^{2+}, Cd^{2+}, (Pb^{2+}), Sn^{2+} and Sn^{4+}, Sb^{3+} and Sb^{5+}

SEPARATION OF THE GROUP II CATIONS

The second set of cations to be removed from a general mixture includes those ions which form insoluble sulfides in the presence of H_2S under highly acidic conditions. These ions make up Group II in the standard qualitative analysis scheme. They precipitate as sulfides at pH 0.5 and are separated from the remaining cations. Since seven different metals form cations which fall into Group II, the procedure for analysis of the ions in this Group is relatively complicated.

PROPERTIES OF THE GROUP II CATIONS

Cu^{2+}. Although copper can exist in the $+1$ oxidation state, copper(I), or cuprous, compounds are relatively rarely encountered. The solution chemistry of copper, for the most part, is that of the copper(II), cupric, Cu^{2+} ion. Most copper(II) salts exist as hydrates in the solid state and are either blue or green. Copper(II) complexes are very common; the dark blue $Cu(NH_3)_4^{2+}$ complex ion is often used to test for the presence of copper(II) ion. Copper(II) is fairly easily reduced to copper(I) or to the metal by the more active metals, such as iron or zinc, or by other strong reducing species, including the hydrosulfite, or dithionite, ion, $S_2O_4^{2-}$:

$$Cu^{2+}(aq) + S_2O_4^{2-}(aq) + 2\ H_2O \rightarrow Cu(s) + 2\ SO_3^{2-}(aq) + 4\ H^+(aq)$$

Bi^{3+}. Bismuth ordinarily occurs in the $+3$ state, but in some strong oxidizing agents it exists as bismuth(V). Bismuth salts are nearly all colorless; they are insoluble in water, but moderately soluble in strong acids, particularly HCl. Bismuth does not tend to form many complex ions. It is reduced to the metal by strong reducing agents, including Sn(II) in basic solution:

$$2\ Bi(OH)_3(s) + 3\ Sn(OH)_4^{2-}(aq) \rightarrow 2\ Bi(s) + 3\ Sn(OH)_6^{2-}(aq)$$

If a solution of a bismuth salt in HCl is added to a large volume of water, the bismuth will precipitate as a white basic salt. In the absence of antimony salts, which behave similarly, this reaction serves as a definitive test for the presence of bismuth ion:

$$Bi^{3+}(aq) + Cl^-(aq) + H_2O \rightarrow BiOCl(s) + 2\ H^+(aq)$$

Cd^{2+}. Cadmium ordinarily exists in its compounds in the $+2$ state. Its salts are usually colorless, and many of them are soluble in water. Cadmium ion forms several complexes, including those with ammonia, halide ions, and cyanide ions. Cadmium is not readily reduced to the metal, and has about the same reduction potential as iron. Cadmium sulfide, CdS, has a characteristic yellow color and is the most soluble of the Group II sulfides. At high concentrations of Cl^- ion,

CdS may not precipitate with the other Group II sulfides because of the stability of the $CdCl_4^{2-}$ complex ion.

Hg^{2+}.   Mercury(II) compounds are much more common than those of Hg(I); the chloride, bromide, cyanide, and acetate are among the mercuric salts that are soluble in water. Mercury(II) salts in water are often only very slightly ionized. The Hg^{2+} ion forms several complexes, including HgS_2^{2-} and $HgCl_4^{2-}$. Mercury(II) sulfide, HgS, is black and is the least soluble of all known sulfides. Mercury(II) is readily reduced to either mercury(I) or to metallic mercury. The deposit of bright mercury on a piece of copper is a common test for the presence of mercury ions in solution. Tin(II) chloride, $SnCl_2$, in acid solution reduces Hg(II) compounds to white Hg_2Cl_2 and/or black metallic mercury.

Sn^{2+} and Sn^{4+}.   Tin in its compounds exists in either the +2, stannous, or the +4, stannic, state. Tin(II), particularly in basic solution, is a good reducing agent. In either state tin forms many complex ions; in solution, tin is usually in the form of a complex anion, such as $SnCl_6^{4-}$ or $Sn(OH)_4^{2-}$. Most tin compounds are colorless; they are essentially all insoluble in water, but will dissolve in either basic or acidic media. Tin(IV) can be reduced to tin(II) by active metals such as iron or aluminum in HCl solution. Tin metal will dissolve in 6 M HCl.

Sb^{3+} and Sb^{5+}.   Antimony salts, like those of tin, are typically insoluble in water. The most common salt of antimony, $SbCl_3$, dissolves readily in moderately strong HCl, forming $SbCl_6^{3-}$, and in NaOH solution, where the antimony species present is the $Sb(OH)_4^-$ complex ion. Sb(III) compounds are much more common than those of Sb(V), the latter being good oxidizing agents. Antimony sulfides are soluble in hot NaOH solution containing S^{2-} ion and in 6 M HCl, as is tin(IV) sulfide. Antimony is reduced to the metal by good reducing agents, such as aluminum in acid solution.

In Table 32.1 we have listed some of the common solubility properties of the Group II cations.

TABLE 32.1   SOLUBILITY PROPERTIES OF THE GROUP II CATIONS

	Cu^{2+}	Bi^{3+}	Cd^{2+}
Cl^-	S	A (white)	S
OH^-	C, A (blue)	A (white)	C, A (white)
SO_4^{2-}	S	A (white)	S
CrO_4^{2-}	C, A (brown)	A (yellow)	C, A (white)
CO_3^{2-}, PO_4^{3-}	C, A (blue)	A (white)	C, A (white)
S^{2-}	O (black)	O (dk brown)	A, O (yellow)
Complexes	NH_3 (blue)	Cl^-	NH_3, Cl^-

	Hg^{2+}	Sn^{2+}, Sn^{4+}	Sb^{3+}
Cl^-	S	C, A (white)	C, A (white)
OH^-	A (yellow)	C, A (white)	C, A (white)
SO_4^{2-}	S	C, A (white)	C, A (white)
CrO_4^{2-}	S^- (red)	S	C, A (yellow)
CO_3^{2-}, PO_4^{3-}	A (red)	C, A (white)	C, A (white)
S^{2-}	O^+, B (black)	A, C, B (tan)	A, C, B (red-orange)
Complexes	Cl^-, S^{2-}	OH^-	OH^-

Key:
S	soluble in water, >0.1 mole/liter
S^-	slightly soluble in water, ~0.01 mole/liter
HW	soluble in hot water
A	soluble in acid (6 M HCl or other non-precipitating, nonoxidizing acid)
I	insoluble in any common solvent

A^+	soluble in 12 M HCl
B	soluble in hot 6 M NaOH containing S^{2-} ion
O	soluble in hot 6 M HNO_3
O^+	soluble in hot aqua regia
C	soluble in solution containing a good complexing ligand
D	unstable, decomposes

GENERAL SCHEME OF ANALYSIS

In the standard scheme for analysis of the Group II cations, following separation of the sulfides, the solid is treated with 1 M NaOH solution, which dissolves the sulfides of tin and antimony and so allows their separation from the rest of the cations in the Group.

The sulfides that were unaffected by NaOH will, with the exception of any HgS, dissolve in hot 6 M HNO_3. The solution obtained is treated with ammonia; in excess NH_3, both Cu^{2+} and Cd^{2+} form complexes, while Bi^{3+} precipitates as the hydroxide, as will any Pb^{2+} that is present. Copper is identified by the dark blue color of its complex ion and by its reduction to the metal with hydrosulfite ion. Any cadmium in the solution is detected by precipitation of yellow CdS following the removal of the copper ion.

The precipitate containing $Bi(OH)_3$ is dissolved in HCl. Bismuth is then identified by precipitation as BiOCl and by reduction to the metal with stannous ion. Although lead is not likely to be present in appreciable amounts, its presence can be confirmed by precipitation, first as the chloride and then as the sulfate. The precipitate of HgS that remains after the other sulfides have dissolved in HNO_3 is dissolved with aqua regia. The presence of Hg(II) is established by reactions with copper metal and with stannous chloride solution.

The solution containing tin and antimony is made acidic, thus reprecipitating the sulfides. These are then dissolved in acid. Half of the acidic sample is treated with aluminum metal, which reduces tin to the $+2$ state and antimony to the metal. Tin is then identified by the reaction of Sn(II) with $HgCl_2$ solution. In the other part of the sample, antimony is detected by precipitation as peach-colored Sb_2OS_2.

PROCEDURE FOR ANALYSIS OF GROUP II CATIONS

WEAR YOUR SAFETY GLASSES WHILE PERFORMING THIS EXPERIMENT

Unless directed otherwise, you may assume that a 10-ml sample contains the equivalent of about 1 ml of 0.1 M solutions of the nitrate or chloride salts of one or more of the Group II cations. If you are working with a general unknown, your sample is the HCl solution decanted from the Group I precipitate and may contain ions from Groups II, III, and IV.

Step 1. Pour 5 ml of your Group II sample or your general unknown into a 50-ml beaker. Add 0.5 ml 3% H_2O_2, and carefully boil the solution down to a volume of about 2 ml.

Step 2. Swirl the liquid around in the beaker to dissolve any salts that may have crystallized, and then pour the solution into a test tube. Add 6 M NaOH a little at a time, until the pH becomes 0.5. (The way to accomplish this is discussed in the comments on procedure following this section.) When the pH has been properly established, add 1 ml 1 M thioacetamide and stir.

Heat the test tube in the boiling-water bath for at least five minutes. **CAUTION:** Small amounts of H_2S will be liberated. This gas is toxic, so avoid inhaling it unnecessarily. If any Group II ions are present, a precipitate will form; typically, its color will be initially light, gradually darkening, and finally becoming black. Continue to heat the test tube for at least two minutes after the color has stopped changing. Cool the test tube under the water tap and let stand for a minute or so. Centrifuge out the precipitate and decant the solution into a test tube. Add 1 ml 1 M NH_4Cl and 1 ml water to precipitate and put it aside.

Check the pH of the decanted solution. If it is too low (too acid), add 0.5 M $NaC_2H_3O_2$, drop by drop with stirring, until the pH is again 0.5. A brown or yellow precipitate may form during this adjustment, indicating that not all of the Group II cations precipitated the first time. In any event, add 0.5 ml 1 M thioacetamide to the decanted solution at pH 0.5, and heat for three minutes in the boiling water bath. Cool the test tube under the water tap and let stand for a minute. Centrifuge out any precipitate, and decant the liquid, which should be saved only if it may contain ions from groups to be studied later. Add 1 ml 1 M NH_4Cl and 1 ml water to the precipitate, stir, and pour the slurry into the test tube containing the first portion of Group II precipitate. Stir well, centrifuge, and decant, discarding the liquid. Wash the precipitate once again with 3 ml water. Centrifuge, and discard the wash liquid.

Under the conditions described here, Cd^{2+} may not precipitate if $[Cl^-]$ is high. If CdS does not form, then cadmium will carry over into Group III. If your unknown contains only the Group II cations, you should force the precipitation of CdS by adding 6 M NH_3 to the solution remaining after removal of the first batch of Group II precipitate; add the NH_3 until the solution is basic to litmus. Then add 0.5 ml 1 M thioacetamide and proceed, starting from the middle of the preceding paragraph.

Step 3. To the precipitate from Step 2 add 2 ml 1 M NaOH. Heat in the water bath, with stirring, for two minutes. Any SnS_2 or Sb_2S_3 should dissolve. The residue will typically be dark and may contain CuS, Bi_2S_3, PbS, CdS, and HgS. Centrifuge and decant the yellow liquid into a test tube (Label 3). Wash the precipitate once with 2 ml water and 1 ml 1 M NaOH and once with 3 ml water. Stir, centrifuge, and decant, discarding the wash each time.

Step 4. To the precipitate from Step 3, add 2 ml 6 M HNO_3. Heat in the boiling-water bath. Most of the reaction will occur within about a minute, as some of the sulfides dissolve and sulfur is formed. There may be a substantial amount of dark residue, which is mainly HgS and free sulfur. Continue heating until no further reaction appears to occur, at least two minutes after the initial changes. Centrifuge and decant the solution, which may contain Cu^{2+}, Bi^{3+}, Cd^{2+}, and Pb^{2+}, into a test tube. Wash the dark residue with 2 ml water, centrifuge, and discard the wash. Then add 1 ml water to the residue and put it aside (Label 4).

Step 5. To the solution from Step 4, add 6 M NH_3 until the solution is basic to litmus. Then add 0.5 ml more and stir. If copper is present, the solution will turn blue. A white precipitate in the solution is indicative of bismuth (or, possibly, lead). Centrifuge and decant the solution, which may contain $Cu(NH_3)_4^{2+}$ and $Cd(NH_3)_4^{2+}$, into a test tube. Wash the precipitate with 1 ml water and 0.5 ml 6 M NH_3. Stir, centrifuge, and discard the wash.

Step 6. To the precipitate from Step 5, add 0.5 ml 6 M HCl and 0.5 ml water. Stir to dissolve any $Bi(OH)_3$ that is present. A white insoluble residue may contain lead. Centrifuge and decant the solution into a test tube. Wash the residue with 1 ml water and 0.5 ml 6 M HCl. Centrifuge and discard the wash.

Step 7. *Confirmation of the presence of bismuth.* Add 2 or 3 drops of the decantate from Step 6 to 300 ml water in a beaker. A white cloudiness caused by precipitation of BiOCl appears if the sample contains bismuth. To the rest of the decantate, add 6 M NaOH until it is definitely basic; a white precipitate is $Bi(OH)_3$. To the mixture add 2 drops 0.1 M $SnCl_2$ and stir; if bismuth is present, it will be reduced to black metallic bismuth.

Step 8. *Confirmation of the presence of lead.* To the white precipitate from Step 6, add 1 ml 0.5 M $NaC_2H_3O_2$, sodium acetate. Heat gently if necessary to dissolve any lead-containing salts. Add 1 ml water and 1 ml 6 M H_2SO_4. A white precipitate of $PbSO_4$ establishes the presence of lead.

Step 9. *Confirmation of the presence of copper.* If the solution from Step 5 is blue, copper must be present, since the color is characteristic of the $Cu(NH_3)_4^{2+}$ ion. To further confirm copper(II), add about 0.3 g of solid $Na_2S_2O_4$, sodium hydrosulfite, to the blue solution. It should quickly decolorize as copper(II) is reduced to copper(I). Put the test tube in the boiling-water bath, where further reduction to reddish or black copper metal will occur. After about a minute, centrifuge out the solid and decant the liquid into a test tube.

Step 10. *Confirmation of the presence of cadmium.* To the decantate from Step 9, or the decantate from Step 5 if that solution was colorless, add 1 ml 1 M thioacetamide and heat in the water bath. Precipitation of yellow CdS will occur within a few minutes if cadmium is present.

Step 11. Pour off the water from the precipitate from Step 4 and add 1 ml 6 M HCl and 1 ml 6 M HNO_3. Put the test tube in the water bath. The dark precipitate, which is mainly HgS, should dissolve in a minute or two, leaving some insoluble sulfur residue. Pour the contents of the tube into a 50-ml beaker and boil gently for about a minute to drive out any remaining H_2S. Add 2 ml water to the solution and stir.

Step 12. *Confirmation of the presence of mercury.* Pour half of the liquid from Step 11 into a test tube. Immerse a 5-mm length of heavy copper wire in the liquid for a minute; a shiny deposit of liquid mercury on the wire will confirm the presence of mercury. To the other half of the liquid from Step 11, in a test tube, add 1 or 2 drops 0.1 M $SnCl_2$; a white, somewhat glossy precipitate of Hg_2Cl_2 again proves the presence of mercury.

Step 13. To the decantate from Step 3 add 6 M HCl drop by drop until the mixture becomes acidic (pH ~0.5, green color on methyl violet). The precipitate that forms may contain the sulfides of Sb(III) and Sn(IV). Stir well for half a minute, centrifuge, and discard the liquid.

Step 14. To the precipitate from Step 13, add 2 ml 6 M HCl. Stir and transfer the mixture to a 50-ml beaker. Boil the liquid gently for one minute to drive out H_2S and to dissolve the sulfides. A black residue is mainly HgS. Add 1 ml 6 M HCl and then pour the solution into a test tube. Centrifuge out any solid residue and transfer the liquid to a test tube. Discard the solid.

Step 15. *Confirmation of the presence of tin.* Pour half of the solution from Step 14 in a test tube and add a 1-cm length of 24-gauge aluminum wire. Heat the test tube in the water bath to promote reaction of the Al and production of H_2. In this reducing medium, any tin present will be converted to Sn^{2+} and any antimony to the metal, which will appear as black specks. Heat for two minutes after all the wire has reacted. Centrifuge out any solid and decant the liquid into a test tube. To the liquid add a drop or two of 0.1 M $HgCl_2$. A white or gray cloudiness, produced as Hg_2Cl_2 slowly forms, establishes the presence of tin.

Step 16. *Confirmation of the presence of antimony.* To the other half of the solution from Step 14, add 6 M NH_3 until the solution is basic to litmus. Any precipitate that forms may be ignored; it will not interfere, but does indicate that either tin or antimony must be present. Add 6 M $HC_2H_3O_2$ until the mixture becomes acidic to litmus; add 0.5 ml more. Add a little (approximately 0.4 g) solid sodium thiosulfate, $Na_2S_2O_3$, and put the test tube in the boiling-water bath for a few minutes. If the mixture contains antimony, a peach-colored precipitate of Sb_2OS_2 will form.

COMMENTS ON PROCEDURE FOR ANALYSIS OF GROUP II CATIONS

Step 1. Here the solution is concentrated, and all ions are brought to their higher states of oxidation. Tin(II) and antimony(III), if present, are oxidized to tin(IV) and antimony(V). This is important, particularly in the case of tin, because unless SnS_2 is the precipitated sulfide, tin will not behave as it should in Step 3.

Step 2. At this point all the Group II sulfides are precipitated with hydrogen sulfide. The reaction that occurs is in all cases analogous to that for copper:

$$Cu^{2+}(aq) + H_2S(aq) \rightarrow CuS(s) + 2 H^+(aq)$$

Since the solubility products of the Group II sulfides are very low, they precipitate in the presence of even extremely low concentrations of S^{2-} ion. The precipitation is carried out at a pH of 0.5, where $[S^{2-}]$ is only about 1×10^{-21} M. The values of the solubility products of some common metallic sulfides are as follows:

Group II		Group III	
HgS	1×10^{-52}	CoS	1×10^{-20}
CuS	1×10^{-35}	NiS	1×10^{-19}
CdS	1×10^{-26}	ZnS	1×10^{-20}
PbS	1×10^{-27}	FeS	6×10^{-17}

Under the conditions of the precipitation, the metallic cations are all about 0.01 M; therefore, $[M^{2+}][S^{2-}] = (1 \times 10^{-2})(1 \times 10^{-21}) \approx 1 \times 10^{-23}$. This means that those sulfides in the left column, $K_{sp} < 10^{-23}$, will precipitate, whereas those in the right column, $K_{sp} \geq 10^{-23}$, will not. This difference in sulfide solubilities is the basis of the separation of the Group II cations (left column) from Group III cations (right column).

Clearly, if the separation is to be effective, the $[H^+]$ must be properly set before precipitation is carried out, since $[S^{2-}]$ is controlled by the pH of the solution. The pH adjustment is probably most easily accomplished with an acid-base indicator. We find that methyl violet is quite suitable. On a piece of filter paper, put about 10 drops of methyl violet indicator, making ten spots about $\frac{3}{4}$ inch in diameter. Let these dry in the air. Make up a solution of HCl about 0.3 M in H^+ ion by diluting 10 ml of a 1 M HCl solution with 20 ml of water and stirring. Put a drop of this solution on a spot of indicator. The blue-green color is that of a solution of the desired pH. Add NaOH, drop by drop, to your cation solution, with stirring, until the color of the indicator, when touched with a drop of solution from your stirring rod, matches that of the standard spot. (Methyl violet is violet at a pH of 2 and yellow at a pH of 0). If you go past the desired pH (too blue a spot), add a drop or two of 6 M HCl.

The source of H_2S in our procedure is thioacetamide, CH_3CSNH_2. This organic compound hydrolyzes when heated in acidic or basic solutions, producing H_2S; in acidic systems the reaction is:

$$CH_3CSNH_2(aq) + 2\ H_2O + H^+(aq) \rightarrow CH_3COOH(aq) + NH_4^+(aq) + H_2S(aq)$$

Generating H_2S in small amounts by this reaction is advantageous because the gas is both bad-smelling and highly toxic. Also, it is produced slowly, which tends to allow the formation of more compact sulfide precipitates.

As you proceed to adjust the pH you will probably find that a precipitate forms. This precipitate consists of basic salts or hydroxides of tin, bismuth, and antimony and will cause no trouble. You may find, however, that accurate pH adjustment is initially difficult, since some of the cations in Group II may interact with the indicator. This again is not serious, since the readjustment of pH following the first precipitation is considerably easier to carry out. It is important that you fix the final pH properly, since too acid a solution (too green an indicator) will not allow precipitation of CdS (yellow), while in too basic a solution (too blue an indicator), sulfides from Group III, particular ZnS (white), will tend to come down.

As we have mentioned, if the chloride ion concentration is too high, CdS will not precipitate even if you adjust the pH perfectly. Since Group II unknowns often contain fairly concentrated HCl to keep tin and antimony in solution, this problem with cadmium is frequently present. In a general unknown, it is less likely, and the problem there is not serious anyway, since cadmium can be identified very nicely in Group III. If your unknown contains only Group II cations, it is advisable to ensure precipitation of CdS by making the solution basic with NH_3 following the initial sulfide precipitation.

Step 3. In this step the sulfides of copper, bismuth, cadmium, and mercury are separated from those of tin and antimony. The former do not dissolve in basic solution, whereas the latter do, according to reactions analogous to that observed with tin:

$$SnS_2(s) + 6\ OH^-(aq) \rightleftarrows Sn(OH)_6^{2-}(aq) + 2\ S^{2-}(aq)$$

Since Sn(II) sulfide would not dissolve, we oxidize tin in Step 1. It is likely that tin and antimony also dissolve as sulfide complexes in this step, but this causes no difficulty. Most of the HgS remains in the sulfide residue.

Step 4. The sulfides of copper, bismuth, and cadmium dissolve readily in hot 6 M HNO_3, a typical reaction being:

$$Bi_2S_3(s) + 8\ H^+(aq) + 2\ NO_3^-(aq) \rightarrow 2\ Bi^{3+}(aq) + 3\ S(s) + 2\ NO(g) + 4\ H_2O$$

Any HgS present is not affected by the nitric acid and remains as a black residue.

Step 5. The blue copper ammonia complex ion will form at this point if the solution contains Cu^{2+} ion. A white precipitate simultaneously formed is highly indicative of bismuth, although if lead is present it will also come down. Cadmium and copper form stable ammonia complex ions, whereas bismuth and lead do not; typically:

$$Cd^{2+}(aq) + 4\ NH_3(aq) \rightleftarrows Cd(NH_3)_4^{2+}(aq)$$

$$Bi^{3+}(aq) + 3\ NH_3(aq) + 3\ H_2O \rightleftarrows Bi(OH)_3(s) + 3\ NH_4^+(aq)$$

The blue color of the solution in Step 5 or Step 9 confirms the presence of copper.

Step 7. There are two very good confirmatory tests for bismuth. Simply pouring a drop or two of the acidic solution of Bi^{3+} into water produces a characteristic white cloudiness owing to the formation of BiOCl. Alternatively, if you make the bismuth solution basic with NaOH and add a drop of $SnCl_2$ solution, an oxidation-reduction reaction occurs, in which black metallic bismuth is formed.

Step 8. In all probability you do not have appreciable amounts of lead appearing in Group II, because the chloride precipitation in Group I is quite effective. Any lead salt you do have will be soluble in sodium acetate solution; addition of sulfuric acid will precipitate $PbSO_4$. Bismuth will not interfere with this test.

Step 9. Here copper is reduced with sodium hydrosulfite, $Na_2S_2O_4$, also called sodium dithionite; addition of the solid to the cold solution causes reduction of copper to the Cu(I) state, and on warming or standing, brown or black metallic copper is produced. Any other cations forming dark sulfides are also reduced, so that only bright yellow CdS is produced when the final solution is treated with sulfide in Step 10.

Step 11. Mercuric sulfide is readily soluble in aqua regia. The solvent used is a rather mild aqua regia:

$$3\ HgS(s) + 12\ Cl^-(aq) + 2\ NO_3^-(aq) + 8\ H^+(aq) \rightarrow$$
$$3\ HgCl_4^{2-}(aq) + 2\ NO(g) + 4\ H_2O + 3\ S(s)$$

Any sulfur formed does not interfere with the tests for mercury.

Step 12. There are two good confirmatory tests for the mercuric ion. The simplest is to put a drop of the solution from Step 11 on a copper penny or to dip a piece of copper wire in the solution. If mercury is present, it quickly deposits as a shiny metal on the copper:

$$HgCl_2(aq) + Cu(s) \rightarrow Hg(l) + Cu^{2+}(aq) + 2\ Cl^-(aq)$$

Probably the most common test is to add a drop or two of $SnCl_2$ solution to the solution being examined; if mercury(II) is present, it is reduced to the mercurous state or to the black metal:

$$2\ HgCl_2(aq) + Sn^{2+}(aq) \rightarrow Hg_2Cl_2(s) + Sn^{4+}(aq) + 2\ Cl^-(aq)$$

Step 13. When the solution from Step 3 is acidified, the hydroxo complexes of tin and antimony are destroyed and the sulfides reprecipitate; reactions like the following one occur:

$$Sn(OH)_6{}^{2-}(aq) + 2\ S^{2-}(aq) + 6\ H^+(aq) \rightarrow SnS_2(s) + 6\ H_2O$$

Step 14. The sulfides of tin and antimony dissolve readily in 6 M HCl at 100°C, with the formation of chloro complex ions. Mercuric sulfide is not soluble in this medium, so a black residue may be observed.

Step 15. Tin is reduced to the +2 state by Al in the presence of acid. The reduction is probably accomplished more by the H_2 gas produced than by the aluminum metal. Sn(II) may be reduced by Al to the metal, but will redissolve in the strongly acidic solution. Any antimony present will be converted to the metal, as will any traces of bismuth. The confirmatory test is again the reaction of $HgCl_2$ with Sn(II).

Step 16. Indications of the presence of antimony appear several times in the procedure. If tin is present, it will not precipitate on addition of the thiosulfate solution. The color of the Sb_2OS_2 is very similar to that of Sb_2S_3.

LABORATORY ASSIGNMENTS

Perform one or more of the following, as directed by your instructor:

1. Make up a sample containing about 1 ml of 0.1 M solutions of the nitrate or chloride salts of each of the cations in Group II. Your instructor will tell you whether you should include Pb^{2+} in the sample. Go through the standard procedure for analysis of the Group II cations, comparing your observations with those that are given. Then obtain a Group II unknown from your instructor and analyze it according to the procedure. On a Group II flow chart, indicate your observations and conclusions regarding the composition of the unknown. Submit the completed chart to your instructor.

2. You will be given an unknown that may contain no more than two cations, both chosen from Group II. You will be told which two cations may be present. Develop as simple a scheme as you can to analyze a solution that might contain one, or both, or none of those ions. Use your scheme to analyze the unknown. Report the results you obtain, along with the procedure you used, to your instructor. Your instructor will then assign you another pair of Group II cations to consider, and an unknown to analyze. Your grade will depend on correct analyses of the unknowns and the number of different analyses you are able to complete in the time allowed. Some possible pairs of cations that may be assigned are the following:

Cu^{2+}, Hg^{2+}	Bi^{3+}, Pb^{2+}	Cd^{2+}, Sb^{3+}
Cd^{2+}, Sn^{4+}	Bi^{3+}, Sb^{3+}	Pb^{2+}, Cu^{2+}

3. Use the analysis scheme from Problem 3 of the Advance Study Assignment to determine the Group II cation present in one or more unknown solutions.

Outline of Procedure for Analysis of Group II Cations

Ions possibly present:

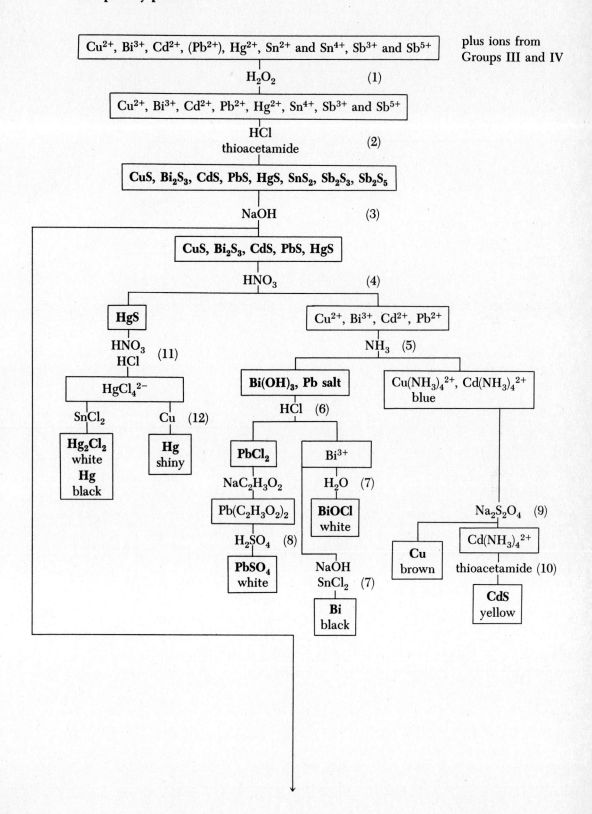

Cu^{2+}, Bi^{3+}, Cd^{2+}, (Pb^{2+}), Hg^{2+}, Sn^{2+} and Sn^{4+}, Sb^{3+} and Sb^{5+} plus ions from Groups III and IV

H_2O_2 (1)

Cu^{2+}, Bi^{3+}, Cd^{2+}, Pb^{2+}, Hg^{2+}, Sn^{4+}, Sb^{3+} and Sb^{5+}

HCl
thioacetamide   (2)

CuS, Bi_2S_3, CdS, PbS, HgS, SnS_2, Sb_2S_3, Sb_2S_5

NaOH   (3)

CuS, Bi_2S_3, CdS, PbS, HgS

HNO_3 (4)

HgS Cu^{2+}, Bi^{3+}, Cd^{2+}, Pb^{2+}

HNO_3 NH_3 (5)
HCl   (11)

$HgCl_4{}^{2-}$ **Bi(OH)₃, Pb salt** $Cu(NH_3)_4{}^{2+}$, $Cd(NH_3)_4{}^{2+}$
                                                blue

$SnCl_2$ Cu   (12)   HCl   (6)

Hg₂Cl₂   **Hg**   **PbCl₂** Bi^{3+}
white   shiny
Hg $NaC_2H_3O_2$ H_2O (7)
black

                        $Pb(C_2H_3O_2)_2$ **BiOCl** $Na_2S_2O_4$ (9)
                                           white
                        H_2SO_4 (8)   **Cu** $Cd(NH_3)_4{}^{2+}$
                                                        brown
                        PbSO₄   NaOH   thioacetamide (10)
                        white $SnCl_2$ (7)
                                                              CdS
                                          Bi   yellow
                                          black

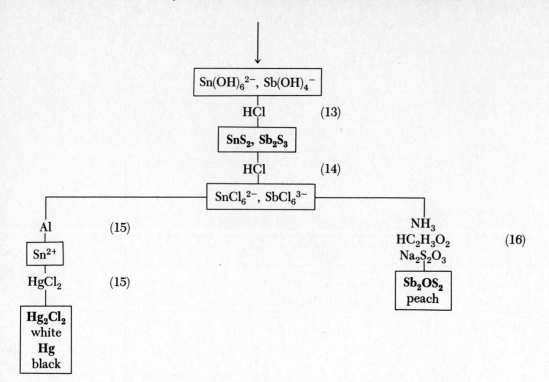

OBSERVATIONS: Analysis of Group II Cations

Flow Chart Showing Behavior of Unknown

ADVANCE STUDY ASSIGNMENT: Analysis of Group II Cations

1. Write balanced net ionic equations for the following reactions:
    a. The confirmatory test for tin (step 15).

    b. A confirmatory test for bismuth (step 7).

    c. The reaction by which CuS is dissolved (step 4).

2. You are given solutions which contain one and only one of the cations in each of the following pairs. For each pair, indicate how you would proceed to determine which cation is present.
    a. Cu^{2+}, Bi^{3+}

    b. Cd^{2+}, Sn^{4+}

    c. Hg^{2+}, Cd^{2+}

    d. Cu^{2+}, Sb^{3+}

3. A solution may contain only one cation from Group II and no other metallic cations. Develop a *simple* scheme to identify the cation. On the back of this sheet, draw a flow chart showing the procedure and observations that establish which cation is present.

4. A solution may contain any of the cations in Group II. The initial sulfide precipitate is unaffected by NaOH. The residue resulting from treatment with NaOH partially dissolves in nitric acid; addition of NH_3 to the acid solution produces a colorless solution and no precipitate. On the basis of these observations, state which of the ions in Group II are present, absent, or still in doubt.

Present _____

Absent _____

In doubt _____

EXPERIMENT

33 • Qualitative Analysis of the Group III Cations: Ni^{2+}, Co^{2+}, Fe^{2+} and Fe^{3+}, Mn^{2+}, Cr^{3+}, Al^{3+}, Zn^{2+}, (Cd^{2+})

SEPARATION OF THE GROUP III CATIONS

In the discussion of the procedure for precipitating the Group II cations, we noted that the relatively low solubility of their sulfides, as compared with those of the Group III cations, allowed for the separation of Group II by sulfide precipitation under moderately acidic conditions. In slightly alkaline sulfide media the Group III cations will precipitate as either sulfides or hydroxides. This property allows one to separate these cations from those in Group IV, none of which precipitate in a buffered sulfide solution maintained at a pH of about 9.

PROPERTIES OF THE GROUP III CATIONS

Ni^{2+}. Nickel salts, like those of most of the other members of Group III, are typically colored. The hydrates are all green, the color of the $Ni(H_2O)_6^{2+}$ complex ion. Nickel forms several complex ions, many of which have characteristic colors; the blue $Ni(NH_3)_6^{2+}$ ion is perhaps the most common of these. Nickel sulfide, NiS, when first precipitated, tends to be colloidal and difficult to settle by centrifuging; it is not very soluble in 6 M HCl even though it cannot be precipitated from acidic solutions. Nickel forms a very characteristic rose-red precipitate with an organic reagent called dimethylglyoxime.

Co^{2+}. Cobalt(II) salts in water solution are characteristically pink, the color of the hydrated cobalt ion, $Co(H_2O)_6^{2+}$. Cobalt(II) forms several complex ions. When heated to boiling, the color of cobalt chloride solutions turns from pink to blue, as the hydrated cobalt ion is converted to species such as $CoCl_3(H_2O)_3^-$. Cobalt sulfide, like NiS, does not dissolve readily in 6 M HCl even when heated. Co(II) reacts with thiocyanate solutions to form a blue complex, $Co(SCN)_4^{2-}$, whose stability is much greater in ethanol than in water. Addition of KNO_2 to solutions of Co^{2+} produces a charcteristic yellow precipitate of $K_3Co(NO_2)_6$:

$$CO^{2+}(aq) + 7NO_2^-(aq) + 3K^+(aq) + 2H^+(aq) \rightarrow K_3Co(NO_2)_6(s) + NO(g) + H_2O$$

Under strongly oxidizing conditions Co(II) can be converted to Co(III), which has a stability that is enhanced in a complex species like $Co(NH_3)_6^{3+}$, or an insoluble substance such as the yellow cobaltinitrite produced in the reaction above.

Mn^{2+}. Manganese exists in its common compounds in perhaps more different oxidation states than any other element, i.e., +2, +3, +4, +6, and +7. The only cation of manganese is Mn^{2+}, which is very pale pink in solution. The common Mn(IV) compound is MnO_2, a dark brown oxide that, like many higher oxides, will dissolve in acids only if a reducing agent like H_2O_2 is present:

$$MnO_2(s) + H_2O_2(aq) + 2 H^+(aq) \rightarrow Mn^{2+}(aq) + O_2(g) + 2 H_2O$$

Potassium permanganate, $KMnO_4$, a deep purple substance that is soluble in water and is a strong oxidizing agent in acid solution, contains manganese in its highest oxidation state, $+7$; the MnO_4^- ion is formed when sodium bismuthate, $NaBiO_3$ is added to Mn^{2+} under acidic conditions:

$$2 \ Mn^{2+}(aq) + 5 \ BiO_3^-(aq) + 14 \ H^+(aq) \rightarrow 2 \ MnO_4^-(aq) + 5 \ Bi^{3+}(aq) + 7 \ H_2O$$

The Mn^{2+} ion, unlike most of the Group III cations, does not form many complexes. The sulfide, MnS, is pink and readily soluble in dilute acids.

Fe^{2+} and Fe^{3+}. Iron in its compounds is ordinarily in the $+2$, ferrous, or the $+3$, ferric, state. The latter is more common, since most ferrous compounds oxidize in the air, particularly if water is present. Iron(II) compounds are usually found as hydrates and are light green. Iron(III) salts are also ordinarily obtained as hydrates and are often yellow or orange. Both Fe^{2+} and Fe^{3+} form many complexes, of which perhaps the most stable are those with cyanide, $Fe(CN)_6^{4-}$ and $Fe(CN)_6^{3-}$. The $FeSCN^{2+}$ ion has a very characteristic deep red color. Metallic iron is a good reducing agent, dissolving readily in 6 M HCl with evolution of hydrogen. Conversion of Fe(II) to Fe(III), or the reverse, is easily accomplished by common oxidizing agents (air, H_2O_2 in acid) or by many reducing agents (H_2S, Sn^{2+}, I^-).

Al^{3+}. Aluminum in its compounds is found essentially only in the $+3$ state. Aluminum salts are typically colorless and soluble in water and in strong acids and bases, but not in ammonia. Aluminum metal is a strong reducing agent and dissolves easily in dilute strong acids and bases, with evolution of hydrogen. Aluminum is typically separated as the hydroxide, which is brought down in an NH_3 solution buffered with NH_4^+. Addition of catechol violet to a slightly acidic solution of Al^{3+} produces a characteristic blue color.

Cr^{3+}. Chromium is ordinarily encountered in the laboratory in either the $+3$ or the $+6$ oxidation state. The common cation is Cr^{3+}, which forms several complexes, essentially all of which are colored; $Cr(H_2O)_6^{3+}$ is reddish-violet in solution. As with aluminum, the precipitate obtained in the Group III precipitation is the hydroxide, $Cr(OH)_3$, rather than the sulfide. Chromium(III) can be oxidized to Cr(VI) with several oxidizing agents (ClO_3^- in 16 M HNO_3, H_2O_2 in 6 M NaOH, ClO^- in 6 M NaOH) and it is in the latter state that it is usually identified. In neutral or basic solutions, Cr(VI) exists as the bright yellow chromate, CrO_4^{2-}, ion. If acid is added to this species, orange dichromate ion, $Cr_2O_7^{2-}$, is produced:

$$2 \ CrO_4^{2-}(aq) + 2 \ H^+(aq) \rightleftharpoons Cr_2O_7^{2-}(aq) + H_2O$$

If H_2O_2 is added to a solution containing $Cr_2O_7^{2-}$ ion, a transitory blue color, due to CrO_5, is observed.

Zn^{2+}. Zinc compounds are typically colorless and hydrated. Most of the salts are soluble in water, dilute acids, strong bases, and ammonia. Zinc ion forms many complexes, the stabilities of two of which explain the solubility of zinc(II) in excess hydroxide and ammonia. Zinc sulfide, ZnS, is white and soluble in dilute strong acids but not in NaOH solutions. Zinc metal is a good reducing agent, reacting with dilute acids to yield hydrogen. In its compounds, zinc occurs ordinarily only in the $+2$ state. In the presence of potassium ferrocyanide, zinc ion forms a characteristic precipitate, $K_2Zn_3[Fe(CN)_6]_2$ (usually blue-green).

In Table 33.1 we have listed the common solubility properties of the Group III cations. In interpreting this and the other solubility tables, you should note that, unless a cation forms an ammonia complex ion, addition of ammonia to that species will typically precipitate its hydroxide or a basic oxide; this will happen with all cations whose hydroxides are insoluble in water and is caused by the fact that NH_3 in solution behaves as a weak base.

TABLE 33.1 SOLUBILITY PROPERTIES OF THE GROUP III CATIONS

	Ni^{2+}	Co^{2+}	Fe^{2+}	Fe^{3+}
Cl^-	S	S	S	S
OH^-	A (green)	A (tan)	A (green)	A (red-brown)
SO_4^{2-}	S	S	S	S
CrO_4^{2-}	S	C, A (brown)	D (brown)	A (tan)
CO_3^{2-}, PO_4^{3-}	C, A (green)	C, A (violet)	A (green)	A (tan)
S^{2-}	A^+, O^+ (black)	A^+, O^+ (black)	A (black)	D (black)
Complexes	NH_3 (blue)	NH_3°	—	—

	Mn^{2+}	Cr^{3+}	Al^{3+}	Zn^{2+}
Cl^-	S	S	S	S
OH^-	A (white)	C, A (green)	C, A (white)	C, A (white)
SO_4^{2-}	S	S	S	S
CrO_4^{2-}	S	A (tan)	C, A (orange)	C, A (yellow)
CO_3^{2-}, PO_4^{3-}	A (white)	A (green)	C, A (white)	C, A (white)
S^{2-}	A (pink)	D (green)	D (white)	A (white)
Complexes	—	OH^-°	OH^-	NH_3, OH^-

° Although cobalt forms ammonia complexes, some of them are insoluble, so that tests for solubility should be made for mixtures of interest. Chromium hydroxide coprecipitated with other hydroxides may be insoluble in 6 M NaOH.

Key:
S   soluble in water, >0.1 mole/liter
S^- slightly soluble in water, ~ 0.01 mole/liter
HW   soluble in hot water
A   soluble in acid (6 M HCl or other non-precipitating, nonoxidizing acid)
I   insoluble in any common solvent

A^+ soluble in 12 M HCl
B   soluble in hot 6 M NaOH containing S^{2-} ion
O   soluble in hot 6 M HNO_3
O^+ soluble in hot aqua regia
C   soluble in solution containing a good complexing ligand
D   unstable, decomposes

GENERAL SCHEME OF ANALYSIS

The qualitative analysis scheme for the Group III cations proceeds rather more easily than that for Group II. Following separation of the Group III precipitate from the solution containing the Group IV cations, we treat the precipitate with 6 M HCl, which puts all the cations except for Ni^{2+} and Co^{2+} into solution. The precipitate of NiS and CoS is dissolved in aqua regia, and the two cations are identified by their reactions with specific reagents. The hydrochloric acid solution of the other cations is made strongly basic and treated with sodium hypochlorite, which oxidizes Cr(III) to Cr(VI), Fe(II) to Fe(III), and Mn(II) to Mn(IV) or Mn(VII). Following treatment with NH_3, the solid phase contains $Fe(OH)_3$, MnO_2, some $Ni(OH)_2$, and possibly some Cd(OH); the solution phase contains CrO_4^{2-}, $Zn(OH)_4^{2-}$, and $Al(OH)_4^-$. The iron, nickel, and cadmium hydroxides are dissolved in H_2SO_4, leaving solid MnO_2, which is then dissolved and tested for the presence of manganese. The Fe^{3+} and Ni^{2+} ions are separated by treatment with ammonia, and identified with specific reagents. If Cd^{2+} is present, it is detected as the sulfide.

The solution of CrO_4^{2-}, $Zn(OH)_4^{2-}$, and $Al(OH)_4^-$ is acidified and then made basic with NH_3, which precipitates aluminum as the hydroxide, allowing its removal and identification. Chromate is precipitated as $BaCrO_4$, and zinc ion is detected with a reagent specific to it.

The tests for identification of the ions in Group III have the advantage that they are relatively free of interferences. In each case, reagents are used that form very characteristic colored precipitates or solutions with the given cations. Some of the identifying species and their properties are as follows:

Ni^{2+}: Nickel dimethylglyoxime, a rose-red precipitate
Co^{2+}: $Co(SCN)_4^{2-}$, a blue complex ion
Fe^{3+}: $FeSCN^{2+}$, a deep red complex ion
Mn^{2+}: MnO_4^-, a deep purple anion

Cr^{3+}: CrO_5, a deep blue unstable peroxide
Al^{3+}: a blue complex with catechol violet
Zn^{2+}: $K_2Zn_3[Fe(CN)_6]_2$, a light green precipitate

PROCEDURE FOR ANALYSIS OF GROUP III CATIONS

WEAR YOUR SAFETY GLASSES WHILE PERFORMING THIS EXPERIMENT

Step 1. Unless directed otherwise, you may assume that 10 ml of your sample contains the equivalent of about 1 ml of 0.1 M solutions of the nitrate or chloride salts of one or more of the Group III cations, plus, possibly, cadmium ion. If you are working with a general unknown, your sample is the solution decanted after separation of the Group II sulfides. Pour 5 ml of your sample, or your general unknown, into a 50-ml beaker and boil the solution gently until the volume is reduced to about 2 ml. Add 1 ml 1 M NH_4Cl and swirl the beaker to dissolve any crystallized salts.

Pour the solution into a test tube and then, drop by drop, with stirring, add 6 M NH_3 until the solution is basic to litmus. Add another 0.5 ml 6 M NH_3. Then add 1 ml 1 M thioacetamide, stir well, and put the test tube in the boiling-water bath for at least five minutes, and at least two minutes after the color of the precipitate stops changing. Stir the mixture occasionally as precipitation proceeds.

Centrifuge out the precipitate and decant the liquid into a test tube. Add a few drops of 1 M thioacetamide to the liquid and put the test tube into the boiling-water bath for a few minutes more to check for complete precipitation of Group III cations. Save the liquid, if necessary, for analysis of Group IV cations. Wash the precipitate twice with 1 ml 1 M NH_4Cl, 2 ml water, and a few drops of 6 M NH_3, stirring well and centrifuging between washes. Discard the wash liquid.

Step 2. To the precipitate from Step 1, which should contain the Group III sulfides or hydroxides, add 1 ml 6 M HCl and 1 ml water. Mix thoroughly and pour the slurry into a 50-ml beaker. Boil the liquid gently for about a minute. The black residue is mainly CoS and NiS. Add 1 ml water, stir, and pour the slurry into a test tube. Centrifuge and decant the liquid, which may contain Al^{3+}, Fe^{2+}, Zn^{2+}, Cr^{3+}, and Mn^{2+}, as well as Ni^{2+} and Cd^{2+}, into a test tube (Label 2). Wash the solid twice with 1 ml 6 M HCl and 1 ml water. Centrifuge and discard the wash liquid.

Step 3. To the precipitate from Step 2, add 0.5 ml 6 M HCl and 0.5 ml 6 M HNO_3. Stir, and put the test tube in the boiling water bath for about a minute. The precipitate should dissolve in a few moments, leaving almost no residue. Add 2 ml water and mix well.

Step 4. *Confirmation of the presence of cobalt.* Pour one third of the solution from Step 3 into a test tube and slowly add about 1 ml of a saturated solution of NH_4SCN, ammonium thiocyanate, in ethyl alcohol, C_2H_5OH. If Co^{2+} is present, a blue solution of $Co(SCN)_4^{2-}$ forms. To the rest of the solution add 6 M NaOH, drop by drop, until a precipitate remains after stirring. Then add 0.5 ml 6 M acetic acid to dissolve the precipitate. To half of this solution add 0.4 g solid KNO_2, potassium nitrite, and stir. A yellow precipitate of $K_3Co(NO_2)_6$, which may form over a period of ten minutes, also confirms the presence of cobalt.

Step 5. *Confirmation of the presence of nickel.* To the other half of the solution from Step 4, add 0.5 ml of dimethylglyoxime reagent. A rose-red precipitate proves the presence of nickel.

Step 6. To the liquid from Step 2, add 6 M NaOH, drop by drop, until the solution is basic, and then add 0.5 ml more. Pour the slurry into a 50-ml beaker and boil gently for two minutes, stirring to minimize bumping. Remove heat, and add 1 ml 1 M NaClO, sodium hypochlorite; swirl the beaker for 30 seconds. Boil the liquid gently, reducing the volume to about 2 ml. Manganese is most likely present if the foam looks purple. Add 0.5 ml 6 M NH_3 and boil for another minute. Transfer the mixture to a test tube and centrifuge out the solid. Decant the liquid, which may contain $Al(OH)_4^-$, $Zn(OH)_4^{2-}$, and CrO_4^{2-}, into a test tube (Label 6). Wash the solid twice with 2 ml water and 0.5 ml 6 M NaOH; centrifuge each time, discarding the wash.

Step 7. To the precipitate from Step 6, which may contain $Fe(OH)_3$, MnO_2, $Ni(OH)_2$, and $Cd(OH)_2$, add 1 ml water and 1 ml 6 M H_2SO_4. Stir and heat the tube for a few minutes in the water bath; centrifuge out any undissolved solid, which should contain essentially only MnO_2. Decant the liquid, which may contain Fe^{3+}, Ni^{2+}, and Cd^{2+}, into a test tube (Label 7). Wash the precipitate with 2 ml water and 1 ml 6 M H_2SO_4; centrifuge and discard the wash.

Step 8. To the precipitate from Step 7, add 1 ml water and 1 ml 6 M H_2SO_4; stir and then add 1 ml of 3 per cent H_2O_2. Put the test tube in the boiling water bath for a minute or so. The precipitate should dissolve readily upon stirring, with possibly a small amount of residue.

Step 9. *Confirmation of the presence of manganese.* Pour 1 ml of the liquid from Step 8 into a test tube and add 1 ml 6 M HNO_3. Add about 0.4 g of solid sodium bismuthate, $NaBiO_3$, with your spatula, and stir well. There should be a little solid bismuthate in excess. Let the mixture stand for a minute and then centrifuge it. If the solution phase is purple, the color is due to MnO_4^- ion and proves the presence of manganese.

Step 10. *Confirmation of the presence of iron.* To one-third of the liquid from Step 7, add 2 ml water and a drop or two of 1 M KSCN, potassium thiocyanate. Formation of a deep red solution of $FeSCN^{2+}$ is a definitive test for iron. To the remaining solution add 6 M NH_3, drop by drop, until the solution is basic, then add 1 ml more. If the sample contains iron, a rust-red precipitate of $Fe(OH)_3$ will form. Centrifuge out the solid and transfer the liquid to a test tube.

Step 11. *Alternative confirmation of the presence of nickel.* The solution from Step 10 may contain $Ni(NH_3)_6^{2+}$ and $Cd(NH_3)_4^{2+}$. To half of that solution add 0.5 ml dimethylglyoxime. Formation of a rose-red precipitate proves the presence of nickel.

Step 12. *Alternative confirmation of the presence of cadmium.* If cadmium failed to precipitate in Group II, it should be in the other half of the solution from Step 10. To that solution add 6 M $HC_2H_3O_2$, acetic acid, until, after mixing, the solution is just acidic to litmus. Then add 1 ml thioacetamide and put the test tube in the water bath. If cadmium is present, a yellow precipitate of CdS will form over a period of several minutes.

Step 13. Returning to the solution from Step 6, add 6 M acetic acid slowly until, after mixing, the solution is definitely acidic to litmus. If necessary, transfer the solution to a 50-ml beaker and boil it to reduce its volume to about 3 ml. Pour the solution into a test tube. Add 6 M NH_3, drop by drop, until the solution is basic to litmus, and then add 0.5 ml in excess. Stir the mixture for a minute or so to bring the system to equilibrium. If aluminum is present, a light, translucent, gelatinous precipitate of $Al(OH)_3$ should be floating in the clear (possibly yellow) solution. Centrifuge out the solid and decant the solution, which may contain CrO_4^{2-} and $Zn(NH_3)_4^{2+}$, into a test tube.

Step 14. *Confirmation of the presence of aluminum.* Wash the precipitate from Step 13 with 3 ml water once or twice, while warming the test tube in the water bath and stirring well. Centrifuge and discard the wash. Dissolve the precipitate in 2 drops 6 M $HC_2H_3O_2$, acetic acid. Add 3 ml water and 2 drops catechol violet reagent and stir. If Al^{3+} is present, the solution will turn blue.

Step 15. If the solution from Step 13 is yellow, chromium is probably present; if it is colorless, chromium is absent. If you suspect that chromium is there, add 0.5 ml 1 M $BaCl_2$. In the presence of chromium, you obtain a very finely divided yellow precipitate of $BaCrO_4$, which may be mixed with a white precipitate of $BaSO_4$. Put the test tube in the boiling-water bath for a few minutes; then centrifuge out the solid and decant the solution into a test tube. Wash the precipitate with 2 ml water; centrifuge and discard the wash.

Step 16. *Confirmation of the presence of chromium.* To the precipitate from Step 15, add 0.5 ml 6 M HNO_3 and stir to dissolve the $BaCrO_4$. Add 1 ml water, stir the orange solution and add two drops of 3 per cent H_2O_2. A blue solution, which may fade quite rapidly, is confirmatory evidence for the presence of chromium.

Step 17. *Confirmation of the presence of zinc.* If you tested for the presence of chromium, use the solution from Step 15. If you were sure chromium was absent, use the solution from Step 13. Make the solution slightly acidic to litmus with 6 M HCl, added drop by drop. Add three drops in excess and then add about three drops of 0.2 M $K_4Fe(CN)_6$, potassium ferrocyanide, and stir. If zinc is present, you will obtain a light green precipitate of $K_2Zn_3[Fe(CN)_6]_2$. Centrifuge to make the precipitate more compact for examination. Decant the liquid, which may be discarded. To the precipitate add 5 drops 6 M NaOH; the solid should dissolve readily with stirring, and possibly some heating in the water bath. If there is an insoluble residue, centrifuge it out and decant the liquid into a test tube. To the liquid add 0.5 ml water and then 6 M HCl, drop by drop, until acidic to litmus (about 6 drops). Re-formation of the green precipitate confirms the presence of zinc.

COMMENTS ON PROCEDURE FOR ANALYSIS OF GROUP III CATIONS

Step 1. As noted previously, the Group II sulfides are extremely insoluble and precipitate with H_2S in acid solution. The sulfides of the cations in Group III are more soluble than those in Group II and can be precipitated by sulfide ion in a slightly basic solution. An ammonia-ammonium chloride buffer is typically used for controlling pH; since in the precipitating medium the concentrations of hydroxide and sulfide ions are low, about 10^{-5} M and 10^{-6} M, respectively, hydroxide and sulfide complexes do not form and Group IV cations, if present, do not precipitate. Because of the extreme insolubility of $Al(OH)_3$ and $Cr(OH)_3$, these compounds, rather than their more soluble sulfides, are precipitated.

In the procedure used, the pH is fixed before thioacetamide is added, and some hydroxides may be precipitated, i.e., $Fe(OH)_3$ (rust red), $Fe(OH)_2$ (green), $Cr(OH)_3$ (gray-green), and $Al(OH)_3$ (white). On formation of the sulfides, the color changes. The sulfides have the following colors: FeS (black), CoS (black), MnS (pink), NiS (black), and ZnS (white). Because the sulfide ion reduces iron from the +3 state, FeS, rather than $Fe(OH)_3$ or Fe_2S_3, should be the iron precipitate. If cadmium carried over from Group II, it will precipitate as yellow CdS.

On centrifuging out the precipitate, you may find that the remaining liquid is dark, with some fine particles dispersed in it. The NiS does not settle out as completely as a well-behaving precipitate should and is responsible for the darkness in the solution. Most of the NiS does come down, though; no real problem is created, just perhaps a little question in one's mind.

Step 2. The relative insolubilities of CoS and NiS make it possible to separate these sulfides from the others by treatment with 6 M HCl. Almost no cobalt(II) dissolves, but some nickel(II) does, and one therefore probably obtains a test for nickel both from the precipitate and from the solution.

Step 4. The use of an alcohol solution of NH_4SCN enhances the stability of the cobalt thiocyanate complex ion and makes the test quite sensitive. If you add the thiocyanate slowly, the lower water layer will typically be red, due to traces of iron, and the upper alcohol layer will be blue. The formation of yellow potassium hexanitrocobaltate(III), $K_3[Co(NO_2)_6] \cdot 3H_2O$, is a less sensitive test for cobalt, but it is also satisfactory. In neither test does nickel interfere.

Step 5. The classic dimethylglyoxime test for nickel is hard to beat. Cobalt does form a colored complex ion with dimethylglyoxime, which must not be confused with a precipitate. However, once you have seen the nickel precipitate made by adding dimethylglyoxime to a dilute nickel solution, you will remember how it looks. Dimethylglyoxime has the formula:

$$CH_3—C=N—OH$$
$$CH_3—C=N—OH$$

and two of these molecules attach to each Ni^{2+} ion in the complex.

Step 6. In this step sodium hypochlorite, NaClO, a very strong oxidizing agent and the main component of household bleach, is used to convert Cr(III) to CrO_4^{2-}. Some of the manganese present may be oxidized to permanganate ion, MnO_4^-, which is deep purple in color. Following the oxidation, NH_3 is added to remove excess hypochlorite and to reduce manganese to Mn(IV), where it precipitates as MnO_2. Some of the key reactions are:

$$2\ Cr^{3+}(aq) + 3\ ClO^-(aq) + 10\ OH^-(aq) \rightarrow 2\ CrO_4^{2-}(aq) + 3\ Cl^-(aq) + 5\ H_2O$$
$$3\ ClO^-(aq) + 2\ NH_3(aq) \rightarrow 3\ Cl^-(aq) + N_2(g) + 3\ H_2O$$
$$2\ MnO_4^-(aq) + 2\ NH_3(aq) \rightarrow 2\ MnO_2(s) + N_2(g) + 2\ OH^-(aq) + 2\ H_2O$$

Never acidify hypochlorite solutions, since chlorine or explosive vapors may be produced.

Step 7. Of the three solids possibly present, only the MnO_2 is so weakly basic that it resists solution in acid; some of the manganese does dissolve, but the bulk remains as a brown solid. If the decanted liquid is dark colored, heat it in the water bath for a few minutes. Centrifuge again to remove the MnO_2 that precipitates, and use the decanted, (usually) light tan liquid in Step 10.

Step 8. In acid solution H_2O_2 reduces many higher oxides, including MnO_2, forming more basic oxides; in this case, MnO, or $Mn(OH)_2$, is produced and dissolves very easily in acid.

Step 9. This is the classic test for the presence of manganese. The only interference is by chlorides and other species that can be oxidized, but in excess bismuthate even these finally are all oxidized and the deep purple MnO_4^- ion forms.

Step 10. The test for Fe^{3+} is another one that is hard to miss. Nitric acid also gives a red coloration, which is the reason H_2SO_4 is used in dissolving $Fe(OH)_3$. Precipitation of rust-colored $Fe(OH)_3$ is also a convincing test for iron.

Step 11. We have always obtained a confirmatory reaction for nickel at this point when nickel was present. It has tended to be somewhat weaker than that in Step 4, however.

Step 14. In the test for aluminum, the solution must be slightly acidic. No other cations that are likely to be present interfere with this test.

Step 16. The blue species formed when H_2O_2 is added to the acidified chromate solution is thought to be CrO_5, an unstable peroxide:

$$Cr_2O_7^{2-}(aq) + 4\ H_2O_2(aq) + 2\ H^+(aq) \rightarrow 2\ CrO_5(aq) + 5\ H_2O$$

Step 17. Potassium zinc hexacyanoferrate(II), $K_2Zn_3[Fe(CN)_6]_2$, is nearly white when pure. In this test it is usually light green or blue-green because of contamination with a trace of iron. Under the acidic conditions that prevail, other cations likely to be present do not interfere.

LABORATORY ASSIGNMENTS

Perform one or more of the following, as directed by your instructor:

1. Make up a sample containing about 1 ml of 0.1 M solutions of the nitrate or chloride salts of each of the cations in Group III. Go through the standard procedure for analysis of the Group III cations, comparing your observations with those that are noted. Then obtain a Group III unknown from your instructor and analyze it according to the procedure. On a Group III flow chart,

Outline of Procedure for Analysis of Group III Cations

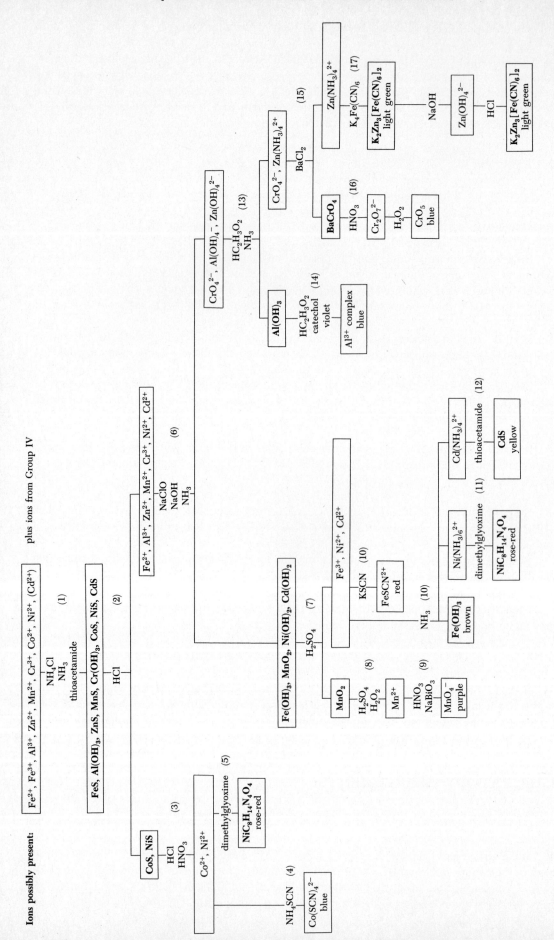

indicate your observations and conclusions regarding the composition of the unknown. Submit the completed chart to your instructor.

2. You will be assigned in advance an unknown that may contain only those Group III cations in one of the following sets. Before coming to the laboratory, develop a procedure by which you can analyze for the ions in your set. Do not include any unnecessary steps; try to make the procedure as simple as possible. Draw a complete flow chart for your scheme, indicating the reagents to be used in each step and the observations that will confirm the presence of each cation. Test your procedure with a sample containing all of the ions in your set, and when you are sure that it works, use your scheme to analyze an unknown furnished to you by your instructor. Record on your flow chart your observations and conclusions for the unknown. Your set of cations will be one of the following:

a. Ni^{2+}, Cr^{3+}, Al^{3+}

b. Co^{2+}, Mn^{2+}, Zn^{2+}

c. Al^{3+}, Zn^{2+}, Ni^{2+}

d. Fe^{3+}, Co^{2+}, Cr^{3+}

e. Cr^{3+}, Mn^{2+}, Co^{2+}

f. Al^{3+}, Co^{2+}, Fe^{3+}

g. Fe^{3+}, Zn^{2+}, Cr^{3+}

h. Ni^{2+}, Mn^{2+}, Cr^{3+}

Name _____ Section _____

OBSERVATIONS: **Analysis of Group III Cations**

Flow Chart Showing Behavior of Unknown

ADVANCE STUDY ASSIGNMENT: Analysis of Group III Cations

1. Write balanced net ionic equations for the following reactions:
   a. The confirmatory test for chromium (step 16).

   b. The reaction by which NiS is dissolved (step 3).

   c. The reaction by which MnS is dissolved (step 8).

   d. The reaction by which Mn^{2+} is converted to MnO_4^- by ClO^- in NaOH solution.

2. You are given solutions that contain one of the cations in each of the following sets. For each set, indicate how you would proceed to determine which cation is present.
   a. Al^{3+} Fe^{3+} Mn^{2+}

   b. Mn^{2+} Co^{2+} Al^{3+}

   c. Pb^{2+} Cr^{3+} Cu^{2+}

3. A solution may contain any of the cations in Group III. Addition of NH_3 and NH_4Cl to the solution gives a reddish precipitate which turns black when H_2S is added. The final precipitate is completely soluble in HCl. Treatment of the HCl solution with NaOCl and NaOH produces a colorless solution and a precipitate that is completely soluble in H_2SO_4. Given this information, state which ions must be present, are absent, or are still in doubt. Give your reasoning for each ion.

Present _____

Absent _____

In doubt _____

EXPERIMENT

34 • Qualitative Analysis of the Group IV Cations: Ba^{2+}, Ca^{2+}, Mg^{2+}, Na^+, K^+, NH_4^+

SEPARATION OF SOME OF THE GROUP IV CATIONS

The ions remaining in solution after the separation of the Group III cations are those of the alkaline earths, the alkali metals, and the ammonium ion. Barium and calcium are separated from the rest of the ions by precipitation of their carbonates under slightly basic conditions. Sodium, potassium, ammonium, and, for the most part, magnesium ions do not precipitate under such conditions. The cations in Group IV have rather less distinctive chemical properties than do the other cations we have studied, so their detection depends, in some cases, more on physical than on chemical behavior.

PROPERTIES OF THE GROUP IV CATIONS

Ba^{2+}. Most of the common barium salts are soluble in water or dilute strong acids and are colorless in solution. The main exception is the sulfate, $BaSO_4$, a finely divided white powder which is essentially insoluble in all common reagents. Barium chromate, $BaCrO_4$, is insoluble in acetic acid but will dissolve in solutions of strong acids. The oxalate and phosphate are soluble in acidic systems at a pH of 3 or less. Barium hydroxide is a strong base and is moderately soluble in water (~ 0.2 moles/liter). Barium exists in its compounds essentially only in the $+2$ state and is not reducible to the metal in aqueous systems.

Ca^{2+}. Calcium salts are typically soluble in water or dilute acids. Like barium, calcium does not form many common complex ions. Although the hydroxide is less soluble than $Ba(OH)_2$, it will not precipitate in ammonia. Calcium oxalate, CaC_2O_4, is white and not appreciably soluble in acetic acid, but it will dissolve in dilute strong acid solutions. Calcium in its compounds occurs in the $+2$ state, and the ion is very difficult to reduce to the metal.

Mg^{2+}. Magnesium salts are all soluble in water or dilute acids. Compounds of magnesium are essentially all white and frequently crystallize as hydrates. Magnesium has the least soluble hydroxide of all the alkaline earths; in the presence of magnesium reagent (4-(p-nitrophenylazo) resorcinol) the hydroxide forms a blue lake. Magnesium oxalate is moderately soluble at pH 5 or below. In an ammonia-ammonium chloride buffer, magnesium ion will form a characteristic white precipitate on addition of Na_2HPO_4 solution:

$$Mg^{2+}(aq) + NH_3(aq) + HPO_4^{2-}(aq) \rightleftharpoons NH_4MgPO_4(s)$$

NH_4^+. Although ammonium ion is not a metallic cation, it forms salts with properties similar to those of alkali metals and is usually included in schemes of qualitative analysis. Ammonium salts are white and soluble in water; if the anion is that of a strong acid, the ammonium salt solution will be slightly acidic. Addition of strong bases to solutions containing ammonium ion causes formation and evolution of NH_3:

$$NH_4^+(aq) + OH^-(aq) \rightleftharpoons NH_3(g) + H_2O$$

The gas can be detected by its odor or by its ability to turn red litmus paper blue. Ammonium salts are much more volatile and unstable when heated than are those of the other cations. The products obtained on heating are often NH_3 and an acid, but N_2 or N_2O may also be produced, depending on the salt.

Na$^+$. Sodium salts are typically water soluble and white. The Na^+ ion in water solution is essentially inert. The hydroxide is very soluble and is the common laboratory source of OH^- ion. Sodium does not form any common complex ions; its salts are frequently obtained as hydrates. It is difficult to detect sodium by precipitation of an insoluble salt; one of the least soluble compounds of sodium is the zinc uranyl acetate, $NaZn(UO_2)_3(C_2H_3O_2)_9$, a yellow substance that can be precipitated from moderately concentrated solutions of Na^+. The most common test for sodium in qualitative analysis is the flame test. In a Bunsen flame, sodium salts give off a very strong characteristic yellow light. Since even traces of Na^+ produce the flame, care is necessary in interpreting the test.

K$^+$. The salts of potassium are similar to those of sodium in their general properties. Potassium hydroxide is very soluble and is a strong base. There are no highly insoluble, easily prepared compounds of potassium; among the least soluble potassium salts is the hexachloroplatinate(IV), K_2PtCl_6. Ordinarily, potassium is detected in qualitative analysis by its characteristic violet flame test, which is much less sensitive than that for sodium.

In Table 34.1 we have listed some of the common solubility properties of the Group IV cations. As a group these cations have by far the most soluble salts; nearly all of their salts will dissolve in very dilute acids.

TABLE 34.1 SOLUBILITY PROPERTIES OF THE GROUP IV CATIONS

	Ba^{2+}	Ca^{2+}	Mg^{2+}	Na$^+$, K$^+$, NH$_4^+$
Cl$^-$	S	S	S	S
OH$^-$	S	S$^-$	A (white)	S
SO$_4^{2-}$	I (white)	S$^-$	S	S
CrO$_4^{2-}$	A (yellow)	S	S	S
CO$_3^{2-}$, PO$_4^{3-}$	A (white)	A (white)	A (white)	S
S^{2-}	S	S	S	S
Complexes	—	—	—	—

Key:
S   soluble in water, >0.1 mole/liter
S$^-$ slightly soluble in water, ~0.01 mole/liter
HW   soluble in hot water
A   soluble in acid (6 M HCl or other non-
     precipitating, nonoxidizing acid)
I   insoluble in any common solvent

A$^+$ soluble in 12 M HCl
B   soluble in hot 6 M NaOH containing S^{2-} ion
O   soluble in hot 6 M HNO$_3$
O$^+$ soluble in hot aqua regia
C   soluble in solution containing a good
     complexing ligand
D   unstable, decomposes

GENERAL SCHEME OF ANALYSIS

Following separation of the Group IV precipitate, containing $BaCO_3$, $CaCO_3$, and possibly some $MgCO_3$, from the soluble cations, Na^+, K^+, and NH_4^+, the precipitate is dissolved in acetic acid. Addition to the resulting solution of chromate, oxalate, and phosphate ions in a succession of steps causes precipitation of $BaCrO_4$, CaC_2O_4, and $MgNH_4PO_4$ and thereby allows a separation of the alkaline earth cations. Specific tests for each cation establish their presence in the solution. Sodium and ammonium ions are tested for in an untreated sample; sodium is identified by its characteristic flame, and ammonium ion by the evolution of NH_3 upon treatment with base. Potassium ion is also detected by a flame test, made either on the original solution or on the liquid decanted from the Group IV precipitate.

PROCEDURE FOR ANALYSIS OF GROUP IV CATIONS

Step 1. If you are working with only the Group IV cations, you may assume, unless told otherwise, that 10 ml of sample contains the equivalent of 1 ml of 0.1 M solutions of the nitrate or chloride salts of one or more of those cations. Pour 5 ml of your Group IV sample into a 50-ml beaker and boil the solution down to a volume of 2 ml. Then add 0.5 ml 6 M HCl, swirl to dissolve any crystals on the walls of the beaker, and transfer the solution to a test tube.

If you are analyzing a general unknown, prepare for Group IV analysis by transferring the solution remaining after removal of the Group III cations to a 50-ml beaker, then boiling it down to a volume of 2 ml. Transfer the liquid to a test tube and centrifuge out any solid matter, which you may discard. Put the liquid back into the beaker, add 1 ml 6 M HCl, and proceed to boil the material essentially to dryness. Transfer the beaker to a hood and carefully heat the dry solid to drive off all the ammonium salts that were added in previous steps. Stop heating the solid when the visible smoke from these salts is no longer evolving. Let the beaker cool, and then add 0.5 ml 6 M HCl and 2 ml water. Warm the solution gently to dissolve the remaining salts; discard any insoluble material after centrifuging it out. Pour the solution into a test tube.

Step 2. To your Group IV sample prepared in Step 1, add 6 M NH_3 until the solution becomes basic; add 0.5 ml more. Then add 1 ml 1 M $(NH_4)_2CO_3$ and stir. Put the test tube in the hot, *but not boiling,* water bath, and without any heating, leave the tube in the bath for two minutes, stirring occasionally. Centrifuge and decant the liquid, which may contain K^+ and Mg^{2+}, into a test tube. Wash the precipitate, which may contain $BaCO_3$, $CaCO_3$, and possibly some $Mg(OH)_2$ or $MgCO_3$, with 2 ml water. Centrifuge out the precipitate, and discard the wash.

Step 3. *Confirmation of the presence of barium.* Add 0.5 ml 6 M acetic acid to the precipitate from Step 2 and stir to dissolve the solid. Add 1 ml water and 2 drops 6 M NH_3 and mix; then add 0.5 ml 1 M K_2CrO_4. A yellow precipitate indicates the presence of barium. Stir for a minute and centrifuge out the solid. Decant the orange liquid, which may contain Ca^{2+} and Mg^{2+}, into a test tube. Wash the solid with 3 ml water; centrifuge, and discard the wash. Dissolve the solid in 0.5 ml 6 M HCl; add 1 ml water. Then add 0.5 ml 6 M H_2SO_4 and stir the solution for 30 seconds. A white precipitate of $BaSO_4$ establishes the presence of barium. Centrifuge to separate the white solid from the orange solution to establish its color and amount.

Step 4. *Confirmation of the presence of calcium.* Add 6 M NH_3 to the decanted orange liquid from Step 3 until it becomes basic; at this point the solution becomes yellow. Add 0.5 ml 1 M $K_2C_2O_4$, potassium oxalate, stir, and let the solution stand for a minute. A white precipitate of $CaC_2O_4 \cdot H_2O$ confirms the presence of calcium. Centrifuge out the white solid and pour the liquid, which may contain Mg^{2+}, into a test tube. Wash the solid with 3 ml of water; centrifuge, and discard the wash.

Dissolve the solid in two drops of 6 M HCl. Perform a flame test several times on the resulting solution. A fleeting orange-red sparkly flame is due to calcium. This color appears before the yellow sodium flame, and it is best observed if you have a small drop of the solution on the test loop; the color lasts only a fraction of a second. Compare your observations here with those obtained with a 0.1 M $Ca(NO_3)_2$ solution.

Step 5. *Confirmation of the presence of magnesium.* To the solution from Step 4, add 0.5 ml 1 M Na_2HPO_4, sodium monohydrogen phosphate. Stir the solution, warm it gently in the water bath, and let it stand for a minute. A white precipitate is highly indicative of the presence of magnesium. Centrifuge out the solid and discard the liquid. Wash the solid with 3 ml of water; centrifuge and discard the wash. Dissolve the solid in a few drops of 6 M HCl. Add 1 ml water and then two or three drops of magnesium reagent, 4-(p-nitrophenylazo)resorcinol. Stir the solution, and then, drop by drop, add 6 M NaOH until the solution is basic. If magnesium is present, a medium blue precipitate of $Mg(OH)_2$ with adsorbed magnesium reagent forms. Centrifuge out the precipitate from the almost colorless solution for better observation of the solid.

Step 6. *Alternative confirmation of the presence of magnesium.* Repeat Step 5 on one half of the solution obtained in Step 2. If magnesium is present in your sample, you will probably detect it in both Step 5 and Step 6; we have found that the amount of blue precipitate is usually greater in Step 6.

Step 7. *Confirmation of the presence of sodium and potassium.* Perform a flame test on 1 ml of the *original* sample using a portion that has not been treated in any way. A strong yellow flame, which persists for several seconds, is confirmatory evidence for the presence of sodium. Since the test is *very* sensitive to traces of sodium, compare the intensity and duration of the flame you obtain with that from a sample of distilled water and that from 0.1 M NaCl solution.

We also perform a flame test on the original sample to detect potassium. The potassium flame is violet and usually lasts for only a moment, perhaps $\frac{1}{2}$ second. The test is much less sensitive than that for sodium, and is best observed with crystals obtained by evaporating 1 ml of sample to near dryness. Look for the potassium flame through one or two thicknesses of blue cobalt glass, which absorbs any sodium emission. The flame from potassium looks reddish-violet through the glass. Compare the flame you observe with that from saturated KCl solution.

Since potassium is much less likely to be a contaminant in your reagents than is sodium, it is also possible to perform a meaningful test for potassium on the other half of the solution from Step 2. Boil that solution down to a volume of only a few drops and carry out the test for potassium on those drops. Because this solution may be quite concentrated in K^+, the test for potassium should be easy to see. In addition, other cations that might interfere with the potassium test in the original sample are not present in the solution remaining after the Group IV precipitation.

Step 8. *Confirmation of the presence of NH_4^+ ion.* Pour the 1 ml of *original* sample used in the flame test for Na^+, or 1 ml of a fresh sample, into a 50-ml beaker. Moisten a piece of red litmus paper and put it on the bottom of a small watch glass. Add 1 ml 6 M NaOH to the sample in the beaker and swirl to stir. Cover the beaker with the watch glass, and then gently heat the solution to the boiling point; do not boil it, and be careful that no liquid solution comes in contact with the litmus paper. If ammonium ion is present, the litmus paper gradually turns blue as it is exposed to the evolved vapors of NH_3. Remove the watch glass and try to smell the ammonia. If only a small amount of ammonium ion is present, the odor test will probably not detect it, whereas the litmus test will.

COMMENTS ON PROCEDURE FOR ANALYSIS OF GROUP IV CATIONS

Step 1. If you have been working with a general unknown, by the time you get to Group IV analysis you have added a substantial number of reagents to the original sample. In particular, ammonium salts have been added; they dilute the Group IV cations and make their precipitation and detection more difficult. Since all ammonium salts are relatively volatile, they are removed from the Group IV metallic cations by simple heating. While subliming the ammonium salts, do not overheat the solid, but keep the temperature at the point at which the salts sublime relatively slowly and smoothly. The nonvolatile salts remaining are easier to dissolve if they have not been heated above the minimal temperature necessary for sublimation.

If you have sufficient general unknown, it is probably best to prepare your sample of Group IV cations by precipitating in one step Groups I, II, and III from a new sample, using Step 1 in the Group III Procedure. This will minimize dilution effects.

Step 2. In this step the carbonates of barium and calcium and possibly the hydroxide or carbonate of magnesium precipitate. The warming of the mixture promotes formation of well-defined precipitates. In the ammonia–ammonium chloride buffer used, magnesium hydroxide or carbonate should not precipitate, though Mg^{2+} does tend to coprecipitate with the other alkaline earths.

Step 3. Neither calcium nor magnesium chromate precipitates under the slightly acidic conditions that prevail here. However, $BaCrO_4$ is essentially quantitatively removed. Its solubility in HCl and reprecipitation of the Ba^{2+} as barium sulfate is definitive evidence for the presence of barium. Lead would not interfere with this test.

Step 4. Neither magnesium oxalate nor barium oxalate from the residual barium present precipitates at this point. A precipitate on addition of oxalate is really definitive evidence for the presence of calcium. Nevertheless, the flame test gives an added degree of confidence that is very reassuring. The calcium flame test lasts only a moment and typically consists of a few red-orange flashes plus a small amount of flame.

In performing a flame test, we use a piece of platinum or chromel wire sealed into the end of a piece of soft glass tubing. Before carrying out the test, make a small loop on the end of the wire and then clean the wire by heating it in a nonluminous Bunsen flame until the flame is colorless; usually the wire will give off a yellow flame initially, but this will gradually disappear as the trace of sodium, which causes the color, vaporizes. Flame tests are best carried out on concentrated solutions or small crystals. To carry out the test, dip the wire into the solution and then put the wire in the Bunsen flame. To pick up a small amount of solid, heat the wire and touch it to the crystals, which will usually adhere sufficiently to be put into the flame.

Steps 5 and 6. The blue lake test for magnesium is excellent. There is interference if cobalt or nickel is present, but that is highly unlikely. You will see the blue $Mg(OH)_2$ precipitate gradually forming as you add the NaOH. Considering that this is the last cation to be precipitated in this scheme, the test is remarkably sensitive.

Step 7. It is very difficult to carry out convincing precipitations of the salts of sodium, potassium, and ammonium ions. The flame tests for sodium and potassium are quite adequate, but they must be done with care. Be sure that your wire is as clean as you can make it before attempting any flame tests. Because your sample will probably show positive results of a sodium test even if almost no sodium is present, be sure to compare your sample with a standard NaCl solution and with distilled water. The potassium test is more easily observed in a partially darkened room.

Step 8. If only a small amount of ammonium salt is present, you probably cannot smell the NH_3 that is evolved. Spattering from the alkaline sample will of course turn the litmus paper blue, but that will not give the smooth, gradual change in color that you get from the NH_3 vapors. Obviously, this test must be done on a sample that has not had either ammonia or ammonium salts added to it.

LABORATORY ASSIGNMENTS

Perform one or more of the following, as directed by your instructor:

1. Make up a sample containing about 1 ml of 0.1 M solutions of the nitrate or chloride salts of each of the Group IV cations. Go through the standard procedure for analysis of Group IV, comparing your observations with those that are described. Obtain a Group IV unknown from your instructor and analyze it according to the procedure. On a Group IV flow chart, indicate your observations and conclusions regarding the composition of the unknown. Submit the completed chart to your instructor.

2. A week in advance, your instructor will assign you a set of seven cations, chosen from the 22 ions in Groups I through IV. Before coming to the laboratory, develop a procedure that you can use to analyze for the ions in your set. Do not include any unnecessary steps, and make the

procedure as simple as possible. Draw a complete flow chart for your scheme, indicating the reagents to be used in each step and the observations that will confirm the presence of each cation. Test your procedure with a sample containing all of the seven ions in your set. When you are sure that your methods work, obtain an unknown for your set of ions from your instructor. Record your observations and conclusions on your flow chart. Your grade will depend upon the analytical results you obtain and upon the validity of the procedure you develop.

Outline of Procedure for Analysis of Group IV Cations

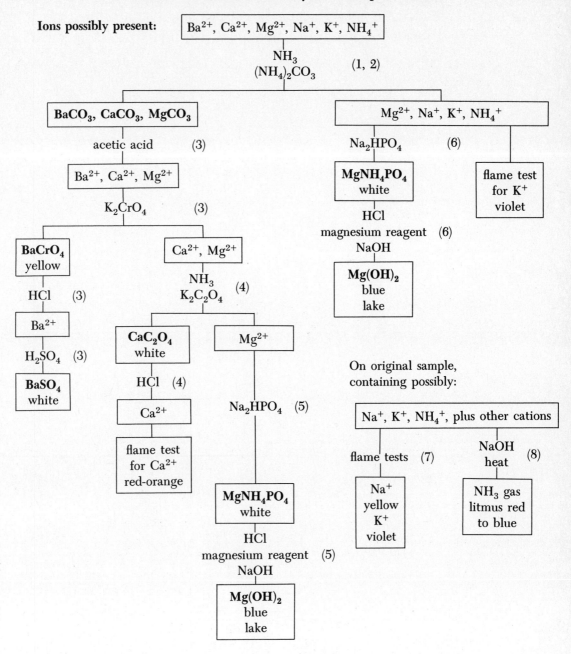

　　　　　　　　Name _____ Section _____

OBSERVATIONS:　Analysis of the Group IV Cations

Flow Chart Showing Behavior of Unknown

ADVANCE STUDY ASSIGNMENT: Analysis of the Group IV Cations

1. Write balanced net ionic equations for the following reactions:
   a. The precipitation of Ca^{2+} in Step 2 in the Group IV Procedure

   b. The final precipitation of Ba^{2+}

   c. The confirmatory test for Mg^{2+} ion

2. For each of the following pairs of ions, name a reagent that, when added in excess to a 0.01 M solution containing both ions, will precipitate the first and leave the second in solution.

   a. Mg^{2+}, Ba^{2+} _____   d. Mg^{2+}, Pb^{2+} _____

   b. Ba^{2+}, Mg^{2+} _____   e. Ag^{+}, Ni^{2+} _____

   c. Al^{3+}, Ca^{2+} _____   f. Cd^{2+}, Zn^{2+} _____

3. Explain why $Ca(OH)_2$ does not precipitate when 6 M NH_3 is added to a solution of Ca^{2+}, although $Bi(OH)_3$ is formed when NH_3 is added to a solution containing Bi^{3+}.

4. An unknown contains four cations, chosen from Groups I through III. The unknown is initially colorless; on addition of 6 M NaOH, a white precipitate is obtained, which dissolves completely in excess reagent. If 6 M NH_3 is added to the original sample, a white precipitate is again obtained, but it is essentially insoluble in excess NH_3. On the basis of this information, can you name the ions in the unknown?

EXPERIMENT

35 • Qualitative Analyis of Anions I The Silver Group: Cl^-, Br^-, I^-, SCN^- The Calcium-Barium-Iron Group: SO_4^{2-}, CrO_4^{2-}, PO_4^{3-}, $C_2O_4^{2-}$

SEPARATION OF THE ANIONS IN THE SILVER GROUP AND THE CALCIUM-BARIUM-IRON GROUP

In anion analysis, the use of groups is more limited than is the case with the cations. In this scheme we separate two groups of anions from the general mixture. Each group separation is usually made from an original sample, rather than sequentially as in cation analysis; this is both convenient and advantageous, since contamination by added reagents is minimized. The Silver Group includes those anions which form insoluble salts with Ag^+ ion in strongly acidic solution. The Calcium-Barium-Iron Group anions form insoluble salts with Ca^{2+}, Ba^{2+}, or Fe^{3+} under slightly acidic conditions. Following separation of the group precipitates, they are dissolved and analyzed on the basis of characteristic properties of the individual anions.

PROPERTIES OF THE ANIONS IN THE SILVER GROUP

Cl^-. Chlorides are among the most common commercially available salts, and, being typically soluble, they are often used as sources of metallic cations in solution. Only the chlorides of the Group I cations, plus the oxychlorides of antimony and bismuth, are insoluble in water. Chloride complexes are quite common and are frequently formed when otherwise insoluble species, such as Sb_2S_3 or $AgCl$, are dissolved in 12 M HCl, the best source of concentrated chloride ion. Chloride ion is not very reactive chemically and is ordinarily detected by the formation of $AgCl$; that salt is insoluble in acids but soluble in ammonia:

$$AgCl(s) + 2\ NH_3(aq) \rightleftharpoons Ag(NH_3)_2^+(aq) + Cl^-(aq)$$

Chloride ion is not easily oxidized to chlorine, Cl_2, but the reaction may occur with strong oxidizing agents like $KMnO_4$ or MnO_2 in very acidic solution:

$$2\ Cl^-(aq) + MnO_2(s) + 4\ H^+(aq) \rightarrow Cl_2(g) + Mn^{2+}(aq) + 2\ H_2O$$

Br^-. Bromide salts resemble the chlorides in most of their general properties. Those metals that have insoluble chlorides have insoluble bromides; as a class, however, the bromides are less soluble than chlorides. $AgBr$ (light tan) is essentially insoluble in 6 M NH_3. Hydrobromic acid, which is a solution of HBr gas in water, is, like HCl, a strong acid. Bromides are more easily oxidized than chlorides, and yield bromine on addition of such oxidizing agents as Cl_2, 15 M HNO_3, and $KMnO_4$ in the presence of strong acids. Ordinarily bromine, Br_2, is detected by its orange color in CCl_3CH_3, 1,1,1-trichloroethane, TCE, in which it is much more soluble than in water.

I^-. Iodides are similar in many ways to the other halide salts, with the exception that they are much more easily oxidized, especially in solution. Iodides are oxidized to I_2 by such reagents as $K_2Cr_2O_7$, KNO_2, H_2O_2, and Cu^{2+} in mildly acidic solutions; stronger oxidizing agents tend to further oxidize I_2 to IO_3^-, which is colorless. The reaction with Cu^{2+} ion is unique for the halides:

$$2\ Cu^{2+}(aq) + 4\ I^-(aq) \rightarrow I_2(aq) + 2\ CuI(s)$$

Silver iodide is a yellow solid and is so insoluble as to be essentially unaffected by 15 M NH_3. Like the other silver halides, AgI can be reduced readily with zinc in acid solution. Iodide ion is often detected by oxidation to I_2, which has a characteristic purple color in CCl_3CH_3.

SCN^-. The thiocyanates resemble the halides in many of their properties, including the insolubility of the silver salt. Like AgCl, AgSCN is soluble in 6 M NH_3. One of the characteristic properties of the SCN^- ion is the deep red color of the $FeSCN^{2+}$ complex ion formed when Fe^{3+} ion is added to solutions of thiocyanates. The thiocyanate ion is easily oxidized, yielding several possible products including sulfate ion, by such species as hot 6 M HNO_3, and $K_2Cr_2O_7$, KNO_2, or H_2O_2 in acid solution. Thiocyanate can also be reduced to H_2S and other species by zinc in dilute H_2SO_4; it is unaffected by zinc in acetic acid.

PROPERTIES OF THE ANIONS IN THE CALCIUM-BARIUM-IRON GROUP

SO_4^{2-}. Sulfate salts are typically soluble in water and are frequently available in chemical stockrooms. Among the common sulfates, only $PbSO_4$ and $BaSO_4$ are insoluble, and, as might be expected, these two salts are often used in sulfate analysis. $PbSO_4$ can be dissolved in strong NaOH solution or in hot concentrated ammonium acetate, in both cases owing to the formation of very stable lead complex ions. Barium sulfate is one of the few common salts that is insoluble in all common solvents. Sulfuric acid, H_2SO_4, from which sulfates can be considered to be derived, is made by dissolving SO_3 in water. Once dissolved, the oxide is not volatile; to drive SO_3 from sulfuric acid one must first boil off most of the water; then, at temperatures above 300°C, SO_3 begins to evolve as a choking gas. Sulfuric acid is one of the strong acids; in the concentrated solution, 18 M H_2SO_4, the acid is also a powerful dehydrating agent and a moderately strong oxidizing agent.

PO_4^{3-}. Phosphate salts are derived from phosphoric acid, H_3PO_4. This weak acid ionizes in three steps, with the successive hydrogen ions being increasingly difficult to remove. The parent acid can be made by dissolving solid P_4O_{10} in water. Most phosphates are insoluble in water but will dissolve in dilute acids. $FePO_4$ is one of the few phosphates that can be precipitated in 1 M $HC_2H_3O_2$ solution. The various phosphate ions are not susceptible to either reduction or oxidation. In hot nitric acid solution, phosphates form a characteristic yellow precipitate of ammonium phosphomolybdate on addition of ammonium molybdate:

$$H_2PO_4^-(aq) + 12\ MoO_4^{2-}(aq) + 3\ NH_4^+(aq) + 22\ H^+(aq) \rightarrow$$

$$(NH_4)_3PO_4 \cdot 12\ MoO_3(s) + 12\ H_2O$$

CrO_4^{2-}. Chromate salts are all colored and are usually yellow or red. Most of these salts are insoluble in water but are soluble in solutions of strong acids. $BaCrO_4$ is precipitated in the Group IV analysis scheme from acetic acid solution, but will dissolve in 6 M HCl. Lead chromate, $PbCrO_4$, is bright yellow and so insoluble that it will not dissolve in dilute strong acids; it will, however, go into solution in 6 M NaOH. In acids, the CrO_4^{2-} ion dimerizes to form orange $Cr_2O_7^{2-}$ ion. Dichromate ion is a good oxidizing agent in strong acid solution, oxidizing such species as SCN^-, I^-, and SO_3^{2-} quite readily, being itself reduced to Cr^{3+}. An unstable blue peroxide, CrO_5, is formed when H_2O_2 is added to strongly acidic dichromate solutions:

$$Cr_2O_7^{2-}(aq) + 4\ H_2O_2(aq) + 2\ H^+(aq) \rightarrow 2\ CrO_5(aq) + 5\ H_2O$$

$C_2O_4^{2-}$. Most of the oxalates are insoluble in water, the exceptions being those of the alkali metals, magnesium, and chromium. The most insoluble oxalate is $CaC_2O_4 \cdot H_2O$, which can be precipitated from acetic acid solution. All of the oxalates will dissolve in 6 M HCl with formation of the weak acid, $H_2C_2O_4$. The oxalate ion is a good complexing agent, forming stable complexes with Sn^{4+}, Fe^{3+}, and Cu^{2+}. The oxalate ion can be slowly oxidized and converted to CO_2 by hot strong oxidizing agents, such as $KMnO_4$ in hot 6 M H_2SO_4:

$$5\ H_2C_2O_4(aq) + 2\ MnO_4^-(aq) + 6\ H^+(aq) \rightarrow 10\ CO_2(g) + 2\ Mn^{2+}(aq) + 8\ H_2O$$

GENERAL SCHEME OF ANALYSIS

Following separation of the silver salts of the halides and SCN^- ion, they are treated with zinc in weakly acidic solution, freeing the anions and reducing the silver ion to the metal. Thiocyanate is identified by formation of the red $FeSCN^{2+}$ complex ion and by oxidation to SO_4^{2-} ion, which is precipitated as $BaSO_4$. Iodide and bromide are oxidized to the elements sequentially by increasingly stronger oxidizing agents, and their presence is established by their characteristic colors in TCE solution. Chloride is precipitated as AgCl after all of the other anions in the group have been oxidized.

In the analysis of the Calcium-Barium-Iron Group, calcium oxalate is precipitated first. Oxalate is identified by its reaction with permanganate. Barium ion is then added to precipitate sulfate and chromate ions. The presence of sulfate is established by the insolubility of $BaSO_4$, and that of chromate by its reaction with peroxide. Finally, phosphate is precipitated by adding Fe^{3+}, and identified, after being redissolved, by formation of its characteristic precipitate with molybdate.

PROCEDURE FOR ANALYSIS OF THE ANIONS IN THE SILVER GROUP

WEAR YOUR SAFETY GLASSES WHILE PERFORMING THIS EXPERIMENT

Unless directed otherwise, you may assume that 10 ml of sample contains about 1 ml of 0.1 M solutions of the sodium salts of the anions in the Silver Group, plus possibly other anions of nonvolatile acids.

Step 1. To 5 ml of your sample add 6 M NH_3 or 6 M HNO_3 until the solution is just basic to litmus. Then add 0.5 ml 6 M HNO_3. Add 1 ml 0.5 M $AgNO_3$ and stir. Centrifuge out the solid, which may contain AgCl, AgBr, AgI, and AgSCN, and check for completeness of precipitation by adding another drop or two of $AgNO_3$. When all the precipitate has been brought down, centrifuge, and decant the liquid, which may be discarded. Wash the solid twice with 3 ml water, discarding the wash.

Step 2. To the precipitate add 2 ml 6 M $HC_2H_3O_2$ and about 0.3 g granular zinc (the volume of the zinc is roughly equal to that of the precipitate). Stir and put the test tube in the boiling-water bath for 10 minutes, stirring occasionally as the reduction of the silver salts proceeds. Cool the test tube under the cold-water tap until the evolution of hydrogen ceases; add 1 ml water, and then centrifuge the mixture, decanting the solution, which may contain Cl^-, Br^-, I^-, and SCN^- ions, into a test tube. Discard the solid residue.

Step 3. *Confirmation of the presence of thiocyanate.* Pour one-third of the liquid from Step 2 into a test tube and add, without stirring, 2 drops of 0.1 M $Fe(NO_3)_3$. If thiocyanate is present, the area where the drops hit the solution will turn a deep rust color owing to formation of $FeSCN^{2+}$.

Since both iodides and acetates will darken slightly on addition of Fe^{3+} ion, confirm the presence of SCN^- by adding 0.5 ml 1 M $BaCl_2$, 0.5 ml 6 M HCl, and 1 ml 3% H_2O_2 to oxidize thiocyanate to SO_4^{2-}. Stir and put the test tube in the water bath for a minute. If thiocyanate is in the sample, a white cloudiness due to the formation of $BaSO_4$ will slowly appear in the yellow liquid. Centrifuge out the solid to establish its color and amount.

Step 4. *Confirmation of the presence of iodide.* To the rest of the solution from Step 2, add 0.5 ml 1,1,1-trichloroethane, CCl_3CH_3(TCE), and 1 ml 1 M KNO_2, potassium nitrite. Stopper the test tube and shake well. If iodide is present, it will be oxidized to I_2 and extracted into the lower TCE layer, where it will have a characteristic purple color.

Step 5. Shake the test tube until the upper water layer is essentially colorless. Let the layers settle and, with a medicine dropper, transfer the upper (water) layer to another test tube. To this, add 0.5 ml 6 M HNO_3 and *slowly*, to avoid excessive foaming, add solid sulfamic acid to remove the excess nitrite. Continue adding solid, with stirring, until there is no more evolution of gas.

Step 6. *Confirmation of the presence of bromide.* To the solution from Step 5, add 0.5 ml TCE and 0.02 M $KMnO_4$, potassium permanganate, a few drops at a time, until the pink color remains for about 15 seconds of stirring. Stopper the test tube and shake. If bromide ion is present, it will be converted to Br_2 by the permanganate and extracted into the TCE layer, where it will have an orange color.

Step 7. Add 5 drops 0.02 M $KMnO_4$, stopper the tube, and shake for 15 seconds to complete oxidation of the bromide ion and extract as much Br_2 as possible into the TCE. Let the layers settle and, using a medicine dropper, transfer the water layer to a 50-ml beaker. Boil the liquid down to a volume of 2 ml, adding 0.02 M $KMnO_4$ dropwise as the boiling proceeds until the liquid takes on a tan cloudiness. Add a few drops of 3% H_2O_2 to dissolve the MnO_2 which has formed, and transfer the solution to a test tube.

Step 8. *Confirmation of the presence of chloride.* To the solution from Step 7, add 0.5 ml 0.1 M $AgNO_3$. If chloride ion is present, you will observe a white cloudiness, which turns to a curdy precipitate of AgCl. To confirm the identification, centrifuge out the solid and decant and discard the liquid. Wash the solid with 3 ml water, centrifuging and discarding the wash. Add 0.5 ml 6 M NH_3 to the solid and stir; if it is AgCl, the solid will dissolve in a few moments. Then add 1 ml 6 M HNO_3, which will cause the AgCl to reprecipitate and prove the presence of chloride.

PROCEDURE FOR THE ANALYSIS OF THE ANIONS IN THE CALCIUM-BARIUM-IRON GROUP

WEAR YOUR SAFETY GLASSES WHILE PERFORMING THIS EXPERIMENT

It may be assumed that 5 ml of sample contains about 1 ml of 0.1 M solutions of the sodium salts of the anions in the Calcium-Barium-Iron Group plus possibly anions of the nonvolatile acids.

Step 1. To 5 ml of your sample add 6 M NaOH or 6 M HCl until the solution is just basic to litmus. Then add 1 ml 6 M acetic acid and mix thoroughly.

Step 2. *Confirmation of the presence of oxalate.* To the solution from Step 1, add 1 ml 1 M $CaCl_2$. Stir for 15 seconds and then put the test tube in the hot-water bath for a few minutes. If a white precipitate forms (usually slowly), it is likely to be $CaC_2O_4 \cdot H_2O$ and therefore indicative of the presence of oxalate. Centrifuge out the solid and decant the liquid into a test tube. Wash the solid twice with 2 ml water, stirring well and warming each time for a minute in the water bath before centrifuging. Discard the wash liquids. To the solid add 2 ml 3 M H_2SO_4. Put the test tube in the water bath, stirring to help dissolve the solid. To the hot liquid add 0.02 M $KMnO_4$, a drop at a time, stirring after each drop. If oxalate is present, it will quickly bleach several drops of the purple solution.

Step 3. *Confirmation of the presence of sulfate.* To half of the liquid from Step 2, add 1 ml 1 M $BaCl_2$. Stir well and centrifuge out any solid that forms, discarding the liquid. Wash the solid, which may be yellow or white and may contain $BaCrO_4$ and $BaSO_4$, with 3 ml water; centrifuge and discard the wash. To the solid add 1 ml 6 M HNO_3 and 1 ml water. Stir and put the test tube in the water bath for about a minute. Any white solid remaining undissolved must be $BaSO_4$ and establishes the presence of sulfate in the sample. Centrifuge out the solid to ascertain its color and amount, and pour the liquid into a test tube.

Step 4. *Confirmation of the presence of chromate.* If the liquid from Step 3 is orange or yellow, chromate is very likely to be present. Cool the test tube under the cold-water tap, and then add 1 ml 3% H_2O_2 to the liquid. A blue coloration, which may fade rapidly, is proof of the presence of chromate.

Step 5. *Confirmation of the presence of phosphate.* To the other half of the liquid from Step 2 add 1 ml 0.1 M Fe $(NO_3)_3$. Formation of a light tan precipitate is highly indicative of the presence of phosphate. Put the test tube in the water bath for a minute to promote coagulation of the solid and then centrifuge, discarding the liquid. Wash the solid with 3 ml water, and heat in the water bath for a minute. Centrifuge and discard the wash. To the solid add 1 ml 6 M HNO_3; the solid should readily dissolve with stirring. To the solution add 1 ml 0.5 M ammonium molybdate and place the test tube in the water bath. If the sample contains phosphate, a fine yellow precipitate of ammonium phosphomolybdate, which may form quite slowly, will be obtained.

COMMENTS ON THE PROCEDURE FOR ANALYSIS OF THE ANIONS IN THE SILVER GROUP

Step 1. The only silver salts that are insoluble in strongly acidic solutions are the silver halides and AgSCN. This precipitation allows a clean and quantitative separation of Cl^-, Br^-, I^-, and SCN^- ions from the other noninterfering anions. (If sulfide ion were present, it would precipitate here as Ag_2S; in the general procedure S^{2-} ion is removed with the other anions of volatile acids before this group separation is made.)

Step 2. Here zinc metal, or hydrogen gas, reduces the silver salts, producing black metallic silver and liberating the anions to the solution, which also ends up containing appreciable amounts of zinc acetate:

$$Zn(s) + 2\ HC_2H_3O_2(aq) \rightarrow Zn^{2+}(aq) + 2\ C_2H_3O_2^{-}(aq) + H_2(g)$$
$$2\ AgCl(s) + H_2(g) + 2\ C_2H_3O_2^{-}(aq) \rightarrow 2\ Ag(s) + 2\ Cl^-(aq) + 2\ HC_2H_3O_2(aq)$$

Step 3. Thiocyanate ion forms a very characteristic deep red complex ion with Fe^{3+}. Unfortunately, both I^- and acetate ions also interact with Fe^{3+} to form much less intensely colored orange species, which conceivably can confuse the thiocyanate test. The confirmation of the presence of SCN^- by its oxidation to sulfate and precipitation as $BaSO_4$ is definitive. Traces of sulfate may cause precipitation of $BaSO_4$ before addition of the peroxide; if this occurs, the precipitate should be centrifuged out and discarded before the solution is treated with peroxide and heated.

Step 4. Nitrite ion oxidizes I^- to I_2 very readily in weak acidic solution. The purple color of I_2 in CCl_3CH_3 is an excellent test for presence of iodide ion:

$$2\ NO_2^-(aq) + 2\ I^-(aq) + 4\ H^+(aq) \rightarrow I_2(aq) + 2\ NO(g) + 2\ H_2O$$

Step 5. Sulfamic acid, NH_2SO_3H, reacts readily with nitrites to produce N_2 gas; we use the reaction here to destroy excess NO_2^-, which would interfere in a later step:

$$NH_2SO_3H(aq) + HNO_2(aq) \rightarrow N_2(g) + HSO_4^-(aq) + H^+(aq) + H_2O$$

Step 6. Permanganate ion under strongly acidic conditions will oxidize bromide ion to bromine, which has an orange color:

$$10\ Br^-(aq) + 2\ MnO_4^-(aq) + 16\ H^+(aq) \rightarrow 5\ Br_2(aq) + 2\ Mn^{2+}(aq) + 8\ H_2O$$

Chloride ion is not appreciably oxidized under the conditions prevailing here.

Step 7. Bromide ion must be completely removed from the solution if the test for chloride ion is to be reliable. In this step residual bromide ion is oxidized to bromine and driven out of the boiling solution.

Step 8. This is the old test for Ag^+, now applied to Cl^-, where it is equally effective.

COMMENTS ON THE PROCEDURE FOR ANALYSIS OF THE ANIONS IN THE CALCIUM-BARIUM-IRON GROUP

Step 2. Under the slightly acidic conditions in the solution, oxalate precipitates essentially quantitatively. Some sulfate ion will tend to come down if its concentration is high.

In the reaction that occurs, MnO_4^- ion oxidizes $C_2O_4^{2-}$ to CO_2 gas and is itself converted to colorless Mn^{2+}. Permanganate ion will decompose slowly in any hot solution, so you should compare the behavior of the test solution with that of 2 ml of hot 6 M H_2SO_4.

Step 3. $BaCrO_4$ and $BaSO_4$ are precipitated here if chromate and sulfate ions are present, but no other barium salts should come down.

There is only one barium salt that is insoluble under the conditions in the HNO_3 solution, and that salt is $BaSO_4$. Traces of sulfate will produce cloudiness, so you may want to compare the amount of precipitate with that obtained from one drop of 0.1 M Na_2SO_4, which contains about 1 mg SO_4^{2-}.

Step 4. The yellow color of CrO_4^{2-}, plus the blue flash of CrO_5, make it hard to miss chromate ion in any solution in which it is present.

Step 5. The precipitate with molybdate will be yellow if phosphate is present. Interfering anions sometimes produce white precipitates.

LABORATORY ASSIGNMENTS

Perform one or more of the following, as directed by your instructor:

1. Make up a solution containing about 1 ml of 0.1 M solutions of the sodium salts of the anions in the Silver Group. Go through the standard procedure for analysis of the Group, comparing your results with those that are described. Then prepare a solution of the anions in the Calcium-Barium-Iron Group and analyze it by the standard procedure. Obtain an unknown, which may contain anions from both of these Groups, and determine its composition. Draw a flow chart for both Groups, showing your observations on the unknown. Note that although the ions within a Group do not react chemically, ions in different Groups may react; in particular here, iodides and thiocyanates are unstable in the presence of chromates in acidic solutions. Consult Appendix III for information regarding behavior of these ions. In dealing with unknowns that exhibit such reactions, you may wish to carry out some experiments with knowns in order to observe directly the changes that occur.

2. Your instructor will assign to you a set of four anions, chosen from the Silver and the Calcium-Barium-Iron Groups. Set up a procedure for the analysis of a sample that may contain only those ions. Draw a complete flow chart for your scheme, indicating all reagents to be used and the observations that will confirm the presence of each anion. Test your procedure with a sample containing all four ions, and when you are sure your approach will work, obtain an unknown for that set of ions from your instructor. Record your observations and conclusions about the unknown on your flow chart and submit it to your instructor.

Outline of Procedure for Analysis of the Anions in the Silver Group

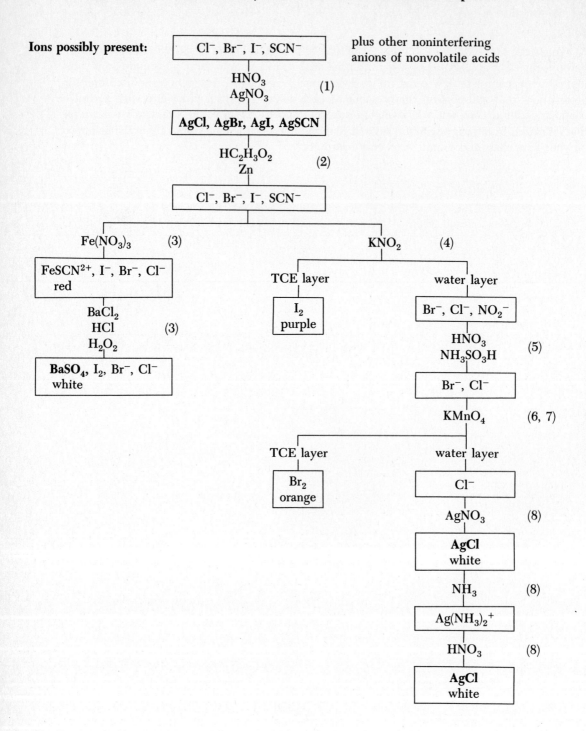

Ions possibly present: Cl^-, Br^-, I^-, SCN^- plus other noninterfering anions of nonvolatile acids

HNO_3
$AgNO_3$ (1)

AgCl, AgBr, AgI, AgSCN

$HC_2H_3O_2$ (2)
Zn

Cl^-, Br^-, I^-, SCN^-

$Fe(NO_3)_3$ (3)

$FeSCN^{2+}$, I^-, Br^-, Cl^-
red

$BaCl_2$
HCl (3)
H_2O_2

BaSO_4, I_2, Br^-, Cl^-
white

KNO_2 (4)

TCE layer

I_2
purple

water layer

Br^-, Cl^-, NO_2^-

HNO_3 (5)
NH_3SO_3H

Br^-, Cl^-

$KMnO_4$ (6, 7)

TCE layer

Br_2
orange

water layer

Cl^-

$AgNO_3$ (8)

AgCl
white

NH_3 (8)

$Ag(NH_3)_2^+$

HNO_3 (8)

AgCl
white

Outline of Procedure for Analysis of Anions in the Calcium-Barium-Iron Group

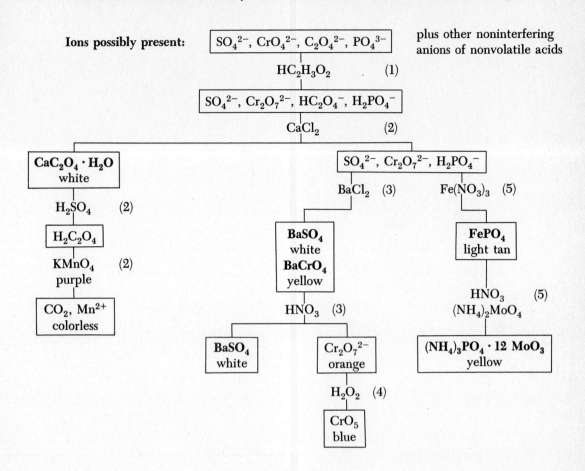

Ions possibly present: SO_4^{2-}, CrO_4^{2-}, $C_2O_4^{2-}$, PO_4^{3-} plus other noninterfering anions of nonvolatile acids

$HC_2H_3O_2$ (1)

SO_4^{2-}, $Cr_2O_7^{2-}$, $HC_2O_4^-$, $H_2PO_4^-$

$CaCl_2$ (2)

$CaC_2O_4 \cdot H_2O$ white

H_2SO_4 (2)

$H_2C_2O_4$

$KMnO_4$ (2) purple

CO_2, Mn^{2+} colorless

SO_4^{2-}, $Cr_2O_7^{2-}$, $H_2PO_4^-$

$BaCl_2$ (3) $Fe(NO_3)_3$ (5)

$BaSO_4$ white $BaCrO_4$ yellow

$FePO_4$ light tan

HNO_3 (3)

HNO_3 (5) $(NH_4)_2MoO_4$

$BaSO_4$ white $Cr_2O_7^{2-}$ orange

$(NH_4)_3PO_4 \cdot 12\ MoO_3$ yellow

H_2O_2 (4)

CrO_5 blue

OBSERVATIONS: Analysis of Anions in the Silver and Calcium-Barium-Iron Groups

Flow Chart Showing Behavior of Unknown

ADVANCE STUDY ASSIGNMENT:　Analysis of the Anions in the Silver Group and
　　　　　　　　　　　　　　　　in the Calcium-Barium-Iron Group

1. Write balanced net ionic equations for the following reactions:
    a. The confirmatory test for bromide ion.

    b. The confirmatory test for oxalate ion.

    c. The confirmatory test for phosphate ion.

2. You are given solutions of the sodium salt of one of the anions in each of the following pairs.
    For each pair, indicate how you would proceed to determine which anion is present.
    a. I^-, Cl^-

    b. SCN^-, Br^-

    c. CrO_4^{2-}, SO_4^{2-}

    d. I^-, PO_4^{3-}

3. A solution may contain only one anion from the Silver Group and no other anions. On the back
    of this sheet, develop a simple scheme to identify the anion. Draw a flow chart showing your
    procedure and the observations that establish which anion is present.

4. A yellow solution may contain any anions in the Calcium-Barium-Iron and the Silver Groups.
    There is no oxidation-reduction reaction on acidification with HNO_3. Addition of $AgNO_3$ to the
    acidified solution yields a yellowish precipitate that is essentially insoluble in 6 M NH_3. There is
    no reaction observed when $BaCl_2$ is added to the same acidified solution. Addition of $CaCl_2$ to the
    original sample produces a white precipitate and a yellow solution. The precipitate is completely
    soluble in acetate acid. On the basis of this information, which anions are present, absent, and still
    in doubt?

    Present　_____

    Absent　_____

    In doubt _____

321

EXPERIMENT

36 • Qualitative Analysis of Anions II
The Soluble Group: NO_3^-, NO_2^-, ClO_3^-, $C_2H_3O_2^-$
The Volatile Acid Group: CO_3^{2-}, SO_3^{2-}, S^{2-}

Although group separations, such as were used in the previous experiment, are possible for some of the common anions, for several others such separations are not feasible. For those ions, spot tests are usually carried out on an original sample in the presence of other anions. In some cases such spot tests are conclusive, but in others there are interferences with the tests if certain other ions are also in the solution. Usually it is possible to remove such interferences prior to carrying out the spot tests, but it certainly is true that in anion analysis it is much more often necessary to devise special procedures for dealing with particular samples than it is in cation analysis.

PROPERTIES OF THE ANIONS IN THE SOLUBLE GROUP

NO_3^-. Since essentially all nitrate salts are soluble in water, they are very commonly used to furnish required cations in laboratory stock solutions. In general, nitrates are quite resistant to reaction in neutral solution, but at high concentrations in highly acidic systems they become good oxidizing agents. Hot 6 M HNO_3 is a strong oxidizing agent, capable of oxidizing such species as I^-, SO_3^{2-}, and CuS. Nitrates can be reduced in 6 M NaOH solution to NH_3 by aluminum or zinc metal. The classic test for nitrates is the "brown ring" test, resulting from the reaction of a concentrated $FeSO_4$ solution with nitrates in strong sulfuric acid; in this test the nitrate ion is reduced to NO and combines with excess Fe(II) to form an unstable brown complex ion, $Fe(NO)^{2+}$:

$$3 \ Fe^{2+}(aq) + NO_3^-(aq) + 4 \ H^+(aq) \rightarrow 3 \ Fe^{3+}(aq) + NO(aq) + 2 \ H_2O$$
$$Fe^{2+}(aq) + NO(aq) \rightarrow Fe(NO)^{2+}(aq)$$

Unfortunately, nitrites, chromates, iodides, and bromides interfere with this test.

NO_2^-. Nearly all nitrite salts are soluble in water. The alkali nitrites, the common laboratory source of the NO_2^- ion, are made by strongly heating the corresponding nitrates. Nitrites can serve as both reducing and oxidizing agents, depending on conditions, and also are fairly good complexing agents. They are unstable in solution, decomposing slowly to nitrates. Since they interfere with many tests for other ions, once they have been detected it is best that they be removed from solution. This can be accomplished by treatment with excess sulfamic acid, NH_2SO_3H, in acidic solution:

$$NO_2^-(aq) + NH_2SO_3H(aq) \rightarrow N_2(g) + SO_4^{2-}(aq) + H^+(aq) + H_2O$$

The reaction is specific to nitrites and produces little, if any, nitrate. Among the oxidizing species we are studying, only nitrite will oxidize I^- ion in cold acetic acid solution. In dilute solutions of strong acids, nitrite ion will oxidize Fe^{2+} to Fe^{3+}; excess Fe^{2+} reacts with the NO produced, forming a brown solution of $Fe(NO)^{2+}$.

ClO_3^-. Chlorate salts are generally soluble in water, but they are not commonly encountered in the laboratory, probably because they are more expensive than the nitrates. Chlorates in acidic solution are moderately good oxidizing agents, and will react, sometimes rather slowly unless heated, with such species as NO_2^-, SO_3^{2-}, and I^-. Hot 6 M HCl is oxidized by chlorates, producing a yellow solution:

$$2 ClO_3^-(aq) + 2 Cl^-(aq) + 4 H^+(aq) \rightarrow 2 ClO_2(g) + Cl_2(g) + 2 H_2O$$

$C_2H_3O_2^-$. Acetates are characteristically soluble in water and are the most common salts of the organic acids. The parent acid, $HC_2H_3O_2$, is moderately volatile and has the odor of vinegar; the odor can be detected above moderately concentrated warm acetate solutions on treatment with 6 M H_2SO_4. In complex mixtures, acetates are best detected by their very characteristic reaction with lanthanum nitrate in the presence of I_3^- ion; iodine adsorbed on the basic lanthanum acetate precipitate gives it a deep blue color. Sulfates and phosphates interfere with this test.

PROPERTIES OF THE ANIONS OF THE VOLATILE ACIDS

CO_3^{2-}. The acid associated with the carbonate ion is carbonic acid, made by dissolving CO_2 gas in water. This acid is quite weak and must dissociate twice to yield the carbonate ion. Carbonic acid cannot be prepared in pure form, and there is some question whether H_2CO_3 molecules exist in solution. Ordinarily, if we were told that a solution was 0.02 M H_2CO_3, we would say that the solution contains 0.02 moles CO_2 per liter. When a carbonate solution is treated with acid, carbonic acid tends to form, and if the solubility of CO_2 is exceeded, CO_2 gas, which is odorless, will effervesce from the system:

$$CO_3^{2-}(aq) + 2 H^+(aq) \rightleftharpoons (H_2CO_3)(aq) \rightleftharpoons CO_2(g) + H_2O$$

Most carbonates are insoluble in water, but dissolve in acids by a reaction similar to the one above, and give off CO_2 in the process. It is possible to prepare hydrogen carbonates, with $NaHCO_3$ being the one most commonly encountered. That salt is also called sodium bicarbonate, or baking soda.

The usual test for carbonates is effervescence on treatment with acid; this will occur with solids and moderately concentrated solutions. The CO_2 driven off is passed by or into a solution of $Ba(OH)_2$, and causes precipitation of white $BaCO_3$. Sulfites and nitrites may interfere, but can usually be removed before the test is made.

SO_3^{2-}. Sulfites can be considered to be derived from sulfurous acid, made by dissolving SO_2 in water. Like carbonic acid, sulfurous acid cannot be isolated in pure form. The acid is weak, but considerably stronger than carbonic acid. Acidified sulfites, whether solid or in solution, evolve SO_2 gas, which can be fairly easily detected by its sharp odor of burning sulfur. Since the gas is not very soluble in water, it is readily driven from a boiling solution. Many sulfites are insoluble in water but soluble in acids; hydrogen sulfites can be prepared and are more soluble than the corresponding sulfites.

The detection of sulfites can be accomplished by the odor test, but in dilute solution it is usually done by oxidation of the sulfite to sulfate (by such reagents as H_2O_2, CrO_4^{2-}, or NO_2^- in acidic solution) and subsequent precipitation as $BaSO_4$. Since sulfites are oxidized by moist air, sulfite solutions will also usually give a positive test for sulfate ion.

S^{2-}. Sulfide ion in solution is most easily detected by the odor of its parent acid, H_2S, which smells like rotten eggs. The acid is very weak, so it exists in appreciable concentrations even in highly alkaline solutions:

$$S^{2-}(aq) + 2 H_2O \rightleftharpoons H_2S(g) + 2 OH^-(aq)$$

Sulfides are typically insoluble; some, like ZnS, will dissolve in 6 M HCl, but others, such as CuS, require oxidation in 6 M HNO_3. H_2S gas could be detected by smell over ZnS when treated with acid, but with CuS the sulfur would end up as either the element or sulfate ion, and there would be very little if any odor of H_2S.

Hydrogen sulfide can be spotted by its unforgettable odor. An alternate test is to pass the gas over a piece of filter paper that has been moistened with lead nitrate solution; formation of shiny black or brown PbS is confirmatory evidence for the presence of sulfide.

As we have noted, anion-anion reactions are common between species that can behave as oxidizing and reducing agents. A summary of the conditions under which reactions occur between such anions is given in Appendix III.

PROCEDURE FOR THE ANALYSIS OF ANIONS IN THE SOLUBLE AND VOLATILE ACID GROUPS

WEAR YOUR SAFETY GLASSES WHILE PERFORMING THIS EXPERIMENT

Unless told otherwise, you may assume that 5 ml of sample contains 1 ml of 0.1 M solutions of the sodium salts of the anions in the Soluble and Volatile Acid Groups plus possibly other nonreacting anions. The tests for the anions in these Groups are all "spot" tests, made on the original sample. Where interferences may occur, the interfering anions are removed before the test is made. The tests are carried out first for those anions for which there are the smallest number of interferences.

Step 1. *Test for the presence of nitrite*. To 1 ml of sample solution add 0.5 ml 6 M H_2SO_4. In a separate test tube, dissolve 0.1 g $FeSO_4 \cdot 7 H_2O$ in 1 ml water. Mix the two solutions. If nitrite is present, you will obtain a dark, greenish brown solution of $Fe(NO)^{2+}$.

Step 2. *Test for the presence of chlorate*. To 1 ml of sample solution, add 0.5 ml 6 M HNO_3 and 2 drops 0.1 M $AgNO_3$. If a precipitate forms, add 2 ml 0.1 M $AgNO_3$, stir, and centrifuge; add a drop of $AgNO_3$ to the liquid to ascertain that all Silver Group anions have precipitated, and that Ag^+ is in excess. To the decanted liquid add 1 ml 1 M KNO_2, potassium nitrite. If chlorate is present, it is reduced by nitrite, and the chloride ion produced reacts with excess Ag^+ to form white AgCl.

Step 3. *Test for the presence of sulfide*. Smell the sample. If it contains sulfide, you should be able to detect the very characteristic and unpleasant odor of H_2S. To further confirm the presence of this ion, pour 1 ml of sample into a 50-ml beaker. On a strip of filter paper, 0.5 × 5 cm, put a drop or two of 0.2 M $Pb(NO_3)_2$; place the filter paper on the bottom of a watch glass. Add 1 ml 6 M H_2SO_4 to the sample and cover the beaker with the watch glass. Heat the liquid gently. If sulfide is present, the paper will turn shiny black or brown as PbS is formed by reaction of Pb^{2+} with the H_2S evolved.

Step 4. *Test for the presence of sulfite*. To 1 ml of sample add 1 ml 6 M HCl and stir. You may be able to detect the sharp odor of SO_2 above the solution. Add 1 ml 1 M $BaCl_2$ to the acidified solution and, after stirring, centrifuge out any precipitate of $BaSO_4$. Decant the clear liquid into a test tube and add 1 ml 3% H_2O_2. If sulfite is present, its oxidation to sulfate causes a new precipitate of $BaSO_4$ to form quickly. See comments for interferences.

Step 5. *Test for the presence of acetate*. To 1 ml of sample add 6 M NH_3 or 6 M HNO_3 until the liquid is just basic to litmus. Add 1 drop 1 M $BaCl_2$. If a precipitate forms, add 1 ml more of the $BaCl_2$ to bring down the anions in the Calcium-Barium-Iron Group. Stir, centrifuge, and decant the clear liquid into a test tube. Add a drop of $BaCl_2$ to make sure that the precipitation was complete. To 1 ml of the liquid add 0.1 M KI_3, drop by drop, until the solution takes on a fairly strong rust color. Add 0.5 ml 0.1 M La(NO$_3$)$_3$ and 3 drops 6 M NH$_3$. Stir and put the test tube in the water bath. In the presence of acetate, the orange gelatinous liquid will gradually darken over a period of several minutes, finally acquiring a dark blue or black color.

Step 6.   *Test for the presence of carbonate.* To 1 ml of sample add 0.5 ml 6 M HCl and stir. Check for the presence of small bubbles effervescing from solution, which is indicative of the presence of carbonate, particularly if the solution is moderately concentrated. Put a stopper *loosely* on the tube and put it in the water bath for a minute. While the tube is warming, draw a little clear, saturated Ba(OH)$_2$ solution into a medicine dropper. Take the test tube out of the water bath, remove the stopper, and put the medicine dropper into the tube so that the tip of the dropper is about 0.5 cm above the liquid surface. Squeeze a drop of Ba(OH)$_2$ so that it hangs from the dropper. If carbonate is present, the CO$_2$ gas evolved from the solution will react with the hydroxide, forming a characteristic "skin" of BaCO$_3$ over the surface of the drop. See comments for interferences.

Step 7.   *Test for the presence of nitrate.* To 1 ml of sample, *carefully* add 2 ml 18 M H$_2$SO$_4$. *CAUTION:* This reagent is caustic. Wash off any spilled reagent immediately with plenty of water. Because this is concentrated acid, add it slowly, mixing all the while. Cool the hot test tube under the water tap. In another test tube dissolve about 0.1 g FeSO$_4$ · 7H$_2$O in 1 ml water. Holding the tube containing the H$_2$SO$_4$ at an angle of about 45 degrees, let 5 drops of the FeSO$_4$ solution from a medicine dropper run down the side of the tube and form a layer over the acid. Let the tube stand for a few minutes, and look for a brown ring which will form at the junction of the two layers if nitrate is present. See comments for interferences.

COMMENTS ON THE PROCEDURE FOR ANALYSIS OF THE ANIONS IN THE SOLUBLE AND VOLATILE ACID GROUPS

Step 1.   Nitrite will react with I$^-$, SCN$^-$, SO$_3{}^{2-}$, S^{2-}, CrO$_4{}^{2-}$, and ClO$_3{}^-$ in weakly acidic solutions, so if any of these ions are present, it is likely that nitrite is not. In this test, some of these ions produce colored solutions, but the Fe(NO)$^{2+}$ complex is very dark and quite distinctive, so there should be no problem with interferences. Given a positive nitrite test, the ions with which it reacts are probably not present. Since they interfere with several tests, nitrites should generally be removed by adding excess sulfamic acid to the acidified sample; nitrite leaves as nitrogen gas, and sulfate ion is introduced into the system. Effervescence on addition of sulfamic acid to a solution is indicative of the presence of nitrite.

Step 2.   Chlorate ion reacts slowly with oxidizable species under acidic conditions. Species that react with nitrite may interfere by removing it, so nitrite should be in excess; products of such reactions will not precipitate with Ag$^+$. Silver ion must of course also be in excess if AgCl is to form.

Step 3.   Sulfide ion is really hard to miss, once you know what H$_2$S smells like.

Step 4.   The test depends on the oxidation of SO$_3{}^{2-}$ ion to SO$_4{}^{2-}$ and its precipitation as BaSO$_4$. Both sulfide and thiocyanate can interfere with this test, since their oxidation products may also be sulfates. The oxidation of sulfide is slow, whereas that of SO$_3{}^{2-}$ and SCN$^-$ is quite rapid. If you can smell SO$_2$ above the acidified solution, that is a good test. Sulfite, if present, will always contain sulfate but thiocyanate will not. Beware of tests for sulfite by BaSO$_4$ precipitation in the presence of SCN$^-$ in solutions that do not contain sulfate ion or evolve SO$_2$ on acidification.

Step 5.   The odor of acetate is not very strong, so it will be hard to detect in dilute solutions. The basic lanthanum acetate precipitate adsorbs I$_2$ much as starch does, resulting in a medium to very dark blue color for the solid, depending on the acetate concentration. If heavy metal cations are present they will precipitate when the sample is made basic. Centrifuge out the solid before adding $BaCl_2$. The addition of $BaCl_2$ in excess removes the interfering anions.

Step 6.   The skin of $BaCO_3$, which forms on the dropper, is quite characteristic, and allows detection of carbonate at concentrations of 0.01 M and higher. There will be noticeable effervescence from solutions that are 0.1 M and higher in CO_3^{2-}. Sulfites and nitrites may interfere with this test; the former give off SO_2, which will form $BaSO_3$ on the drop of $Ba(OH)_2$, and the latter tend to effervesce. If either ion is present, add 0.5 ml 3% H_2O_2 to the neutral solution before you acidify it; put the test tube in the hot-water bath for a few minutes to allow the reaction to be completed. Cool the tube, add the acid, and proceed as before. The test for carbonate is difficult in the presence of sulfite, since any SO_2 that is liberated may interfere. Repeat the test if it is not conclusive, giving the H_2O_2 at least three minutes to complete the oxidation.

Step 7.   The ring is caused by formation of brown $Fe(NO)^{2+}$ ion. Oxidizable or reducible anions, like I^-, Br^-, NO_2^-, and CrO_4^{2-} may interfere with the test. Iodides and bromides produce dark colored solutions, nitrites produce the same ring as nitrates only more so, and chromates form a green ring. Remove iodide and bromide by adding Ag_2SO_4 in 1 M H_2SO_4 in excess. (Make this reagent by mixing 2 ml 0.1 M $AgNO_3$ with 3 drops 6 M NaOH. Centrifuge out the brown Ag_2O that forms and wash the solid with water. Add 2 ml water and 1 ml 3 M H_2SO_4 and stir.) Chromates will precipitate on addition of 1 M $BaCl_2$ in excess to the neutral solution. Nitrites decompose without formation of nitrate on addition of excess sulfamic acid to the slightly acidic sample. After removal of the many interferences, the test should be dependable. However, it would be wise, when interferences are present, to compare your test with that obtained with 0.01 M KNO_3.

LABORATORY ASSIGNMENTS

Perform one or more of the following, as directed by your instructor:

1.   Carry out the standard tests for the identification of the anions in the Soluble and Volatile Acid Groups. For each test, use a solution of the 0.1 M sodium salt of the anion. Repeat each test with a solution made by mixing two drops of the 0.1 M salt solution with 1 ml water. Using the standard tests, analyze an unknown that contains anions from the Soluble and Volatile Acid Groups. Since there are many possible anion-anion reactions and anion interferences for these tests, examine the Table of Anion-Anion Reactions in Appendix III and the interferences mentioned in the Comments on Procedure for this experiment before making your report on the composition of the unknown.

2.   You will be given an unknown that contains the sodium salts of five of the anions studied in this and the previous experiment. The anions are selected so that there will be no anion-anion reactions occurring when the unknown is acidified with a strong acid. There may, however, be anion interferences with given tests. Before carrying out the tests for anions in the Calcium-Barium Group, acidify the sample with 6 M HCl and boil gently for two minutes to remove anions of the volatile acids. If sulfide is present, acidify the sample with 6 M H_2SO_4 and boil for two minutes before testing for anions in the Silver Group. Analyze the unknown for its composition, and check to see that it contains no reacting anions and that all interferences have been considered. In your report, describe the behavior of the sample in each test as well as the anions it contains.

Name _____ Section _____

OBSERVATIONS: Analysis of Anions in the Soluble and Volatile Acid Groups

Test	Stock Solution 0.1 M and 0.01 M	Unknown
Nitrite		
Chlorate		
Sulfide		
Sulfite		
Acetate		
Carbonate		
Nitrate		

ADVANCE STUDY ASSIGNMENT: Analysis of Anions in the Soluble and the Volatile Acid Groups

1. Write balanced net ionic equations for the following reactions:
   a. The confirmatory test for sulfite ion.

   b. The removal of interferences with the carbonate test.

2. You are given solutions that contain the sodium salt of one of the anions in each of the following pairs. For each pair, indicate how you would proceed to determine which ion is present.

   a. SO_3^{2-}, SO_4^{2-}

   b. ClO_3^-, CO_3^{2-}

   c. NO_3^-, CrO_4^{2-}

3. A neutral solution contains only one anion, at a concentration of 0.3 M, chosen from the set of 15 we have studied. You are allowed to use up to five reagents to determine which ion it is. Draw up a complete flow chart on the back of this sheet, showing how you would proceed in performing the analysis.

EXPERIMENT

37 • Identification of a Pure Ionic Compound

So far in our work in qualitative analysis we have dealt with solutions of the ions under consideration, assuming in general that the cations were in the presence of nitrate or chloride ions and that the anions were obtained from solutions of their sodium or potassium salts. It is by no means necessary to limit qualitative analyses to such systems, although they are very useful when one is becoming familiar with the analytical procedures. At this point, we are ready to deal with a rather more realistic problem, namely, the identification of a pure ionic solid substance by determining the cation and anion that it contains.

In principle, we could find the anion and cation by dissolving the sample and then going through the procedures for analysis of the cations in Groups I to IV and the anions in the four groups we have considered. This would be highly inefficient, since we can often use solubility tests to limit the number of possible cations and anions to only a few. We then carry out spot tests for only those ions, using, in most cases, the identification tests in the qual scheme. We have to dissolve the solid in any event, and it turns out that the solubility behavior of the solid is usually highly indicative of the ions it contains.

As possible solvents for the solid, we use the following reagents under both hot and cold conditions:

1. water
2. 6 M H_2SO_4
3. 6 M HCl
4. 6 M HNO_3
5. 6 M NH_3
6. 6 M NaOH

The acids in the list may dissolve a water-insoluble solid by reacting with the anion it contains, either in an acid-base or an oxidation-reduction reaction. Such reactions often produce effervescence or color changes, either in the solution or in the gas that is evolved. The bases in the list may dissolve a water-insoluble solid by reacting with the cation it contains, forming an ammonia or hydroxo complex. Your observations on treating an unknown solid with the six reagents we have listed can be very helpful in the task of finding out which cation and anion are present.

Since it is not obvious how one proceeds to identify cations and anions on the basis of solubility behavior, let us consider a particular example. Let's assume that a white solid unknown showed the following behavior with the reagents on the list, as tested at room temperature and then in a hot-water bath:

1. water: soluble, colorless solution

2. 6 M H_2SO_4: soluble, colorless solution

3. 6 M HCl: soluble, colorless solution

4. 6 M HNO_3: orange solution, orange gas on heating

5. 6 M NH_3: insoluble, white residue

6. 6 M NaOH: insoluble, white residue

Certainly, the most spectacular reaction is the one with HNO_3 when the mixture is heated. As we noted, some anions—in fact most anions—react with strong acidic solutions. Not all the reactions are as dramatic as the one in our example, but most of them are highly characteristic of a particular anion. In Table 37.1 we have listed the behavior to be expected for ten of the anions we studied in qualitative analysis. The behavior, as you can see, is remarkably different for each anion and will usually allow you to eliminate all but one or two anions if there is any indication at all that a reaction with one of the acidic reagents occurred. Our example clearly indicates that the bromide ion is on the scene, with really no other likely candidate.

TABLE 37.1   BEHAVIOR OF IONIC SOLIDS CONTAINING PARTICULAR ANIONS ON TREATMENT WITH ACIDIC REAGENTS

Anion	6 M H_2SO_4	6 M HNO_3	6 M HCl
CO_3^{2-}	Strong effervescence with these reagents(C); odorless, colorless CO_2 gas produced.		
SO_3^{2-}	Effervescence with these reagents(C); colorless SO_2 gas produced; has odor of burning sulfur.		
NO_2^-	Effervescence with these reagents(C); brown NO_2 gas produced; solution is blue with H_2SO_4 and HNO_3; yellow with HCl.		
ClO_3^-	With HCl(H), yellow-green solution, gas evolved with odor of Cl_2.		
I^-	Yellow solution with HNO_3(C); violet I_2 vapor, dark solution with HNO_3(H); red-orange with H_2SO_4(H); yellow solution with HCl(H).		
Br^-	Orange solution, orange Br_2 gas with HNO_3(H).		
SCN^-	Rapid effervescence with HNO_3(H); brown NO_2 evolved; with H_2SO_4(H) and HCl(H), slight effervescence.		
CrO_4^{2-}	Solution obtained with these reagents is orange(C).		
S^{2-}	Reasonably soluble sulfides produce H_2S with all reagents(C). Highly insoluble sulfides in HNO_3(H) yield brown NO_2 gas, plus a light or dark residue.		
$C_2H_3O_2^-$	Slight odor of vinegar produced with all reagents(H); most easily detected by cautiously sniffing stirring rod dipped in hot solution.		

These observations are typical of those observed with reasonably soluble compounds.

Key:
(C) No heating used; stirring for two minutes if no obvious reaction.
(H) Heating for up to five minutes in a water bath.

Having concluded that, in all probability, the anion is Br^-, you could proceed to do the confirmatory test for Br^- in the Silver Group, using a water solution of the unknown. If that test is positive, Br^- is indeed present. If it is not, you'd best check your observations, since something is amiss.

There are seven common anions that will not give a good indication of their presence in the course of the solubility tests. These anions are NO_3^-, Cl^-, SO_4^{2-}, PO_4^{3-}, $C_2O_4^{2-}$, OH^-, and O^{2-}. We will discuss their identification in the Procedure for Analysis.

Once the anion in the unknown has been identified, or at least limited to a small number of possibilities, we go on to cation identification. Here the solubility properties in the different reagents can also be very helpful. To assist you in the interpretation of your observations, we have listed in Appendix II the solubility characteristics of most of the salts and hydroxides of the cations and anions we have investigated; incidentally, oxides often behave in much the same way as hydroxides, but may take a great deal longer to react. The information in Appendix II is much like that in the tables we included in the cation analysis, but now we need to use it in our anlaysis.

Our ionic unknown was found to be soluble in water. In the Appendix the entry S means that the compound at the cation-anion intersection is soluble in water. Our sample must therefore be one of those characterized by an S. Since it's probably a bromide, the cation would be one whose bromide is described by an S. Since the sample and its solutions are not colored, it could not contain a colored cation such as Fe^{3+} or Cu^{2+}. The fact that the solid is soluble in HCl confirms that it cannot contain a Group I cation, such as Ag^+ or Pb^{2+}; neither AgCl or $PbCl_2$ is described with an S. Since the solid is insoluble in 6 M NH_3 and in 6 M NaOH, the cation does not form a soluble complex(C) with either NH_3 or OH^- ion; this eliminates Al^{3+}, Cd^{2+}, and Zn^{2+}, whose hydroxides all have a C as one of the entries. The insolubility in NH_3 and NaOH also means that the hydroxide of the cation in the solid is insoluble, since otherwise the solid would dissolve in the basic solutions. This rules out Na^+, K^+, NH_4^+, whose hydroxides are soluble(S), and possibly Ba^{2+} and Ca^{2+}, which have slightly soluble hydroxides(S^-).

At this point we can limit the cations that might be in the solid to the following four: Mg^{2+}, Ba^{2+}, Ca^{2+}, and Hg^{2+}. Carrying out the Group IV procedure for the first three ions on a water solution of the solid should determine if the cation is in Group IV. Mercury could be identified by using its confirmatory test in Group II on a water solution of the solid.

Since the arguments we used to eliminate all the cations except Mg^{2+}, Ba^{2+}, Ca^{2+}, and Hg^{2+} are fairly complex, you should now, without consulting the previous paragraphs, prove to yourself that Appendix II and the solubility properties of the unknown indeed indicate that the four cations we have listed are the only ones that could exist in the unknown solid. For effective use of Appendix II you need to understand what the data there tell you, so some practice in interpreting that data, which you can get by working out the problems in the Advance Study Assignment, will be useful when you get to analyze your own sample.

PROCEDURE FOR ANALYSIS   WEAR YOUR SAFETY GLASSES WHILE PERFORMING THIS EXPERIMENT

This procedure requires that you have available at least one gram of solid sample. Unless told otherwise, you may assume that the sample consists of a pure ionic compound. The sample should be finely divided, preferably powdered, since powders tend to dissolve more readily than large crystals. If the sample does not appear to be affected reasonably promptly by the reagents in Step 1, it may be helpful to grind it in a mortar, testing first with a tiny portion to make sure it will not explode.

Step 1. *Solubility characteristics*. Determine the solubility behavior of the sample in the reagents listed below, using about 50 mg of solid (enough to cover a 2-mm circle on the end of a small spatula) and about 0.5 ml of reagent in a test tube; test with the reagents listed below:

1. water	3. 6 M HCl	5. 6 M NH_3
2. 6 M H_2SO_4	4. 6 M HNO_3	6. 6 M NaOH

With each reagent, stir the solid with your stirring rod for a minute or two, observing any changes in color of the solid or the solution. If a gas evolves, note its color and odor. Then put the test tube in the boiling-water bath. Note any changes that occur, leaving the tube in the water bath for a few minutes. Add 5 ml water to dilute the system. Stir and, if solid is still present, heat for a few more minutes. Record all your observations.

Compare your observations with those listed in Table 37.1. If your anion appears to be one that is listed in the Table, carry out a spot test for that anion, using a solution of the solid (50 mg in 1 ml water or 1 ml 6 M HNO_3) and the identification test in the qualitative analysis procedure. If the test involves a gas, and the solid is insoluble in water, use 50 mg of the solid. If your sample is insoluble in all of the acidic reagents, proceed to Step 4.

Step 2. *Anion analysis*. At this point you may know which anion is in your unknown. If that is the case, proceed to Step 3. If your observations in Step 1 were inconclusive, or if there were no clear reactions other than that the sample did or did not dissolve, then proceed with this step. Once you have identified the anion in your unknown, go on to Step 3.

Prepare a solution of the sample. Use water, 6 M HCl, or 6 M HNO_3, whichever was the least reactive solvent. Put about 0.1 g of the solid in a test tube, add 0.5 ml of acid if you need it, stir, and then add 5 ml water. Put the test tube in the water bath if the solid does not dissolve readily. Use the solution in the following spot tests:

Test for the presence of sulfate. To 1 ml of solution add 1 ml 6 M HCl and a few drops of 1 M $BaCl_2$ and stir. If sulfate ion is present, a white, finely divided precipitate of $BaSO_4$ will form.

$BaSO_4$ is the only barium compound that will not dissolve in strongly acidic solution, so the test is a good one even with complex mixtures.

Test for the presence of phosphate. To 1 ml of solution add 1 ml 6 M HNO_3. Then add 1 ml 0.5 M $(NH_4)_2MoO_4$, ammonium molybdate, and stir. Warm the test tube in the water bath for a minute or two, and then let it stand in air for up to ten minutes. In the presence of phosphate, a yellow precipitate of ammonium phosphomolybdate will form.

Test for the presence of oxalate. Add 6 M NaOH to 1 ml of solution until it is just basic to litmus. Then add 0.5 ml 6 M acetic acid and 1 ml 1 M $CaCl_2$. Stir and let stand for up to ten minutes. Formation of a white precipitate is highly indicative of the presence of oxalate. Centrifuge out any precipitate and decant the solution, which may be discarded. Wash the solid twice with 4 ml water plus 0.5 ml 6 M acetic acid, centrifuging and discarding the wash each time. To the solid add 1 ml 6 M H_2SO_4 and put the test tube in the hot-water bath to dissolve the solid. To the hot solution add, drop by drop, 0.02 M $KMnO_4$. If oxalate is present it will quickly bleach several drops of the purple solution. In addition, you may observe bubbles of CO_2 produced.

Test for the presence of nitrate. If the anion in your sample is nitrate, the sample will be soluble in water. Carry out the test for nitrate in the Soluble Group, using the water solution you prepared in this Step.

Test for the presence of chloride. If the anion in your sample is chloride, the solid will dissolve in water in many cases, and 6 M HNO_3 in other cases. Make a solution of 50 mg of sample in 0.5 ml 6 M HNO_3 and 5 ml water. To 1 ml of solution add a few drops of 0.1 M $AgNO_3$. If chloride ion is present, you will get a white curdy precipitate of AgCl. Centrifuge out the precipitate and discard the solution. Wash the precipitate with 5 ml water, centrifuge, and discard the wash. To the precipitate add 1 ml 6 M NH_3. The solid should dissolve. Then add 6 M HNO_3 until the solution is acidic. If the solid reprecipitates, the sample contained chloride ion.

If no confirmatory tests for anions were obtained, as carried out in connection with the results of the solubility tests in Step 1 and Table 37.1, or in this step, it is likely that the solid is an oxide or hydroxide. In that case your solid will typically be insoluble in water but soluble in acids, unless of course the acid contains an anion that precipitates with the cation in the sample. If you suspect you have a hydroxide or oxide, add 6 M NaOH to a solution of your sample in 6 M HNO_3. A precipitate at that point is consistent with the presence of hydroxide or oxide but is not proof that either anion is present. Proof of the presence of hydroxide and oxide ultimately depends on your being able to show that the other possible anions are not present.

It is possible that the cation in your sample may interfere with your anion analysis. If this seems to be the case, add 2 M Na_2CO_3 to 1 ml of your sample solution until any effervescence ceases and the solution is definitely basic. Add 0.5 ml more. Centrifuge out the solid precipitate, which will contain your cation. Decant the solution into a test tube, add 6 M H_2SO_4 until evolution of CO_2 ceases, and the solution is slightly acidic. Use this solution with the anion tests.

Step 3. Cation analysis. Go back to your observations in Step 1, this time with the purpose of identifying the cation in your sample. Using Appendix II and the solubility properties of your diluted mixture, eliminate as many cations as possible from consideration. Consider water solubility, solubility in acids, complex-ion formation, and color of precipitates and solutions in making your deductions. By proper interpretation of your observations, you should be able to narrow the list of likely cations to no more than four. In some cases you will be able to identify the cation for sure; in others, there may be two or three possibilities.

To make the solution to use for identification of your cation, add about 50 mg of the unknown solid to 5 ml water and 1 ml 6 M HNO_3, or 6 M HCl, in a 50-ml beaker. Use whichever acid was the better solvent; HNO_3, if they were both effective. Boil the solution gently for two minutes to remove any volatile or reactive anions. Use the solution in identification tests for the cations that were reasonable in light of your solubility tests. In general you should not have to go through more than a few steps in the qualitative analysis procedure; for the most part you can go directly to the confirmation test for the cation in question, adjusting pH if necessary to the proper value before making the test.

If your unknown contains oxalate or phosphate ions, and your solubility tests imply that Ba^{2+}, Ca^{2+}, or Mg^{2+} is the likely cation, the standard Group IV procedure may cause some difficulty. It is probably best to first spot test for Ba^{2+}; this is done by adding two drops 6 M H_2SO_4 to 1 ml of the unknown solution you prepared in this step; a white, finely divided precipitate could only be $BaSO_4$ or $PbSO_4$. Since Pb^{2+} was presumably already eliminated, a precipitate proves the presence of Ba^{2+}. If barium is absent, spot test for magnesium with another 1 ml sample of the unknown solution; make the solution just basic with 6 M NH_3, and carry out the confirmatory test for Mg^{2+} in Group IV. If magnesium is absent, add 1 ml 1 M $K_2C_2O_4$, potassium oxalate, to a third 1-ml sample of the unknown solution; slowly add 6 M NH_3 to the solution until it is just basic to litmus; if calcium ion is present, you should get a white precipitate containing CaC_2O_4. Centrifuge out the solid, wash with 3 ml water, centrifuge again. Confirm the presence of Ca^{2+} with the flame test, as in the Group IV procedure.

Step 4. *Analysis of insoluble solids.* Although most solids will dissolve in at least one of the acidic reagents used in Step 1, there are a few that will not. In some cases, the solid sample you were given may have aged for many years or may have been prepared at a high temperature; under such conditions the sample may dissolve much more slowly than one would expect, so you might give it some extra time in the hot-water bath in Step 1.

If, after all reasonable efforts the solid does not dissolve in any of the acids, it is probably one of the substances listed in Table 37.2. In that table we include some substances that dissolve in either 6 M NH_3 or 6 M NaOH, since such behavior is often helpful in identifying ionic compounds.

TABLE 37.2   SOME COMMON ACID-INSOLUBLE SUBSTANCES

Substance	Properties
AgCl (white) AgSCN (white)	1. Insoluble in 6 M acids and NaOH solutions 2. Dissolve in 6 M NH_3, 1 M $Na_2S_2O_3$, and 12 M HCl
AgBr (light yellow) AgI (yellow)	1. Insoluble in acids and NaOH solutions 2. AgBr dissolves in 17 M NH_3, both AgBr and AgI dissolve in 1 M $Na_2S_2O_3$
Hg_2Cl_2 (white) Hg_2Br_2 (white) Hg_2I_2 (orange)	1. Insoluble in 6 M acids and bases 2. Dissolve in aqua regia 3. Turn dark on addition of 6 M NH_3
$PbSO_4$ (white) $PbCl_2$ (white) $PbBr_2$ (white) PbI_2 (yellow) $PbCrO_4$ (yellow)	1. Insoluble in 6 M acids; $PbCrO_4$ is somewhat soluble in 6 M HNO_3 2. Dissolve in 6 M NaOH and in 3 M $NaC_2H_3O_2$ 3. $PbCl_2$ is reasonably soluble in hot water
$BaSO_4$ (white)	1. Insoluble in 6 M acids and bases and in aqua regia 2. Transposes slowly to $BaCO_3$ in 2 M Na_2CO_3
SnO_2 (white) Al_2O_3 (white)	1. Very resistant to all solvents if solids were subjected to high temperatures
HgS (black)	1. Insoluble in 6 M acids and bases 2. Dissolves in aqua regia
MnO_2 (dark brown)	1. Insoluble in most 6 M acids and bases 2. Dissolves in 6 M acids plus 3% H_2O_2 3. Dissolves in hot 6 M HCl, forming Cl_2 and Mn^{2+}
PbO_2 (dark brown)	1. Insoluble in most 6 M acids and bases 2. Reacts with hot 6 M HCl, forming Cl_2 and white $PbCl_2$

If your sample dissolves only in basic solutions, you would have trouble with many of the confirmatory tests in the qualitative analysis scheme, since most of these tests involve acidic solutions.

Given one of these solids as an unknown, you should be able to limit the possibilities rather quickly on the basis of the properties listed in the table. Solubility in 6 M NH_3 is highly indicative of AgCl or AgSCN. Darkening upon treatment with 6 M NH_3 implies you have a Hg(I) salt. Solubility in 6 M NaOH is characteristic of lead salts. No effect with any acids or bases is only observed with $BaSO_4$, SnO_2 and Al_2O_3, all of which are white, and with HgS, which is black.

Once you have an idea which cation is likely to be present, you may be able to free the anion by reducing the cation to the metal. To 50 mg of solid in a test tube add 1 ml 6 M HCl and, slowly, about 100 mg of solid granulated zinc. Silver-, lead-, and mercury-containing salts will be reduced to the metals by the H_2 gas that is produced, and the anion will go into solution in most cases (S^{2-} and SCN^- may be converted to H_2S and evolved). The solution will also contain Zn^{2+} and excess acid. You can then dilute with water, centrifuge, and test the decantate for likely anions. If the anion is a halide, the reduction is best carried out with 6 M H_2SO_4. The metal can then be identified by treating the residue from the centrifugation with 6 M HCl to dissolve all of the zinc, and then dissolving the metal that remains in 6 M HNO_3. Use the confirmatory test in the qualitative analysis scheme to identify the cation.

Another approach that is sometimes effective is to transpose the insoluble salt to another insoluble salt. This can be done by treating about 50 mg of the solid with 2 ml 2 M Na_2CO_3 in a 50-ml beaker. Boil the mixture gently for five minutes, adding water as necessary to maintain volume. If all goes well the residue will contain your cation as a carbonate, and the solution will contain the anion along with excess CO_3^{2-} and Na^+. Centrifuge out the solid carbonate and decant the solution into a test tube. Dissolve the solid in 1 ml 6 M HNO_3. The solution that is obtained can be used to test for the cation in the unknown. To the decanted solution add 6 M HNO_3 until acidic and all the effervescence is gone. Test that solution for the anion in the unknown.

Analysis of Complex Mixtures

The general problem of analysis of mixtures that contain more than one cation and one anion is considerably more difficult than that of analysis of a pure substance. A prerequisite to mixture analysis is enough experience with pure substances to make their analyses seem quite easy.

The properties of a mixture will ordinarily be a superposition of those of the substances it contains. The procedure we have given for analyzing a pure substance will in general apply to the mixture. You may find that the solubility behavior of the mixture will give you a good idea about two, or even three, anions that are present. Partial solubility in, or reaction with, one or more of the reagents may offer clues about the cations in the mixture. If the sample is water soluble, or soluble in one of the acids, you can use the solution to go through the procedure in Groups I to IV in the cation analysis scheme. It would be best to prepare the solution with HNO_3, if possible, and boil for a few minutes to remove volatile or reactive anions. You can use the water solution, or a solution in any of the acids, to test for anions, paying due regard to possible reactions of the anions with acid, as well as interferences that can result from anion-anion reactions or anion-cation reactions. Some mixtures can be analyzed about as easily as pure substances; others may require all the skills of a very experienced analytical chemist.

LABORATORY ASSIGNMENTS

Perform one or more of the following, as directed by your instructor:

1.   You will be given a salt that contains one cation and one anion. Identify the salt, using the procedure we have described. In your report, indicate the experiments you performed during the analysis that established the cation and the anion that were present.

2. You will be given a solution that contains two cations and two anions in addition to possibly H^+, Na^+, and NO_3^- ions. Determine the ions in the solution. Describe in your report the procedure you used and the results you obtained.

3. You will be given a solid mixture that may include salts, oxides, and hydroxides. The solid will contain three cations and three anions, in addition to possibly Na^+, OH^-, O^{2-}, and NO_3^- ions. Determine the ions present in the solid. In your report, describe completely the procedure you used, including those observations that confirmed the identification of each ion.

Name _____ Section _____

OBSERVATIONS AND REPORT SHEET: Analysis of a Pure Ionic Solid

List your observations in Step 1. Record all the confirmatory tests you performed and the reasoning that led you to identify the unknown as you did.

Anion present _____

Cation present _____

ADVANCE STUDY ASSIGNMENT: Analysis of a Pure Ionic Solid

1.   A blue solid is insoluble in water but dissolves in 6 M HCl, 6 M HNO_3, and 6 M H_2SO_4, with considerable effervescence but no noticeable odor in the gas evolved. The solution produced is green with HCl and blue with HNO_3 and H_2SO_4. In 6 M NH_3, the solid dissolves, forming a dark blue solution. The solid does not appear to be affected by 6 M NaOH. On the basis of this information and the data in Appendix II, identify the solid. State your reasoning.

2.   A white solid is insoluble in all the acidic reagents used in Step 1. It is also insoluble in 6 M NH_3 but does dissolve readily in 6 M NaOH. In hot water it shows no appreciable solubility. Name two substances that might be in the solid. State your reasoning.

3.   A white solid is soluble in water and in all of the acidic reagents used in Step 1; the solutions produced are all colorless, and in no case is there any evidence of effervescence or production of a vapor with an odor. The solid appears to react with 6 M NH_3 to produce a white insoluble substance. In 6 M NaOH the solid dissolves readily. The solution of the solid in 6 M HNO_3 yields a white precipitate on treatment with 1 M $BaCl_2$. Identify the solid.

APPENDIX I

Vapor Pressure of Water

Temperature °C	Pressure mm Hg	Temperature °C	Pressure mm Hg
0	4.6	26	25.2
1	4.9	27	26.7
2	5.3	28	28.3
3	5.7	29	30.0
4	6.1	30	31.8
5	6.5	31	33.7
6	7.0	32	35.7
7	7.5	33	37.7
8	8.0	34	39.9
9	8.6	35	42.2
10	9.2	40	55.3
11	9.8	45	71.9
12	10.5	50	92.5
13	11.2	55	118.0
14	12.0	60	149.4
15	12.8	65	187.5
16	13.6	70	233.7
17	14.5	75	289.1
18	15.5	80	355.1
19	16.5	85	433.6
20	17.5	90	525.8
21	18.7	95	633.9
22	19.8	97	682.1
23	21.1	99	733.2
24	22.4	100	760.0
25	23.8	101	787.6

APPENDIX II

Summary of Solubility Properties of Ions and Solids

	Cl$^-$, Br$^-$ I$^-$, SCN$^-$	SO$_4$$^{2-}$	CrO$_4$$^{2-}$	PO$_4$$^{3-}$	C$_2$O$_4$$^{2-*}$	CO$_3$$^{2-}$
Na$^+$, K$^+$, NH$_4$$^+$	S	S	S	S	S	S
Ba^{2+}	S	I	A	A$^-$	A	A$^-$
Ca^{2+}	S	S$^-$	S	A$^-$	A	A$^-$
Mg^{2+}	S	S	S	A$^-$	A$^-$	A$^-$
Fe^{3+} (yellow)	S$^\circ$	S	A$^-$	A	S	D,A$^-$
Cr^{3+} (blue-gray)	S	S	A$^-$	A	S	A$^-$
Al^{3+}	S	S	A$^-$,C	A,C	A$^-$,C	D,A$^-$,C
Ni^{2+} (green)	S	S	S	A$^-$,C	A,C	A$^-$,C
Co^{2+} (pink)	S	S	A$^-$	A$^-$	A$^-$	A$^-$
Zn^{2+}	S	S	A$^-$,C	A$^-$,C	A$^-$,C	A$^-$,C
Mn^{2+} (pale pink)	S	S	S	A$^-$	A$^-$	A$^-$
Cu^{2+} (blue)	S$^\circ$	S	A$^-$,C	A$^-$,C	A,C	A$^-$,C
Cd^{2+}	S	S	A$^-$,C	A$^-$,C	A,C	A$^-$,C
Bi^{3+}	A	A$^-$	A	A	A	A$^-$
Hg^{2+}	S$^\circ$	S	A	A$^-$	A	A$^-$
Sn^{2+}, Sn^{4+}	A,C	A,C	A,C	A,C	A,C	A,C
Sb^{3+}	A,C	A,C	A,C	A,C	A$^-$,C	A,C
Ag$^+$	C$^\circ$	S$^-$	A,C	A,C	A,C	A$^-$,C
Pb^{2+}	C,HW	C	C	A,C	A,C	A$^-$,C
Hg$_2$$^{2+}$	O$^+$	A	A	A	O	A

	SO$_3^{2-}$	S^{2-}	O^{2-},* OH$^-$,	NO$_3^-$, ClO$_3^-$, C$_2$H$_3$O$_2^-$, NO$_2^-$	Complexes
Na$^+$, K$^+$, NH$_4^+$	S	S	S	S	—
Ba^{2+}	A	S	S$^-$	S	—
Ca^{2+}	A$^-$	D,A$^-$	S$^-$	S	—
Mg^{2+}	S	D,A$^-$	A$^-$	S	—
Fe^{3+} (yellow)	D,S	D,A$^-$	A$^-$	S	—
Cr^{3+} (blue-gray)	S	D,A$^-$	A$^-$	S	*
Al^{3+}	A$^-$,C	D,A$^-$,C	A$^-$,C	S	OH$^-$
Ni^{2+} (green)	A$^-$	O	A$^-$,C	S	NH$_3$
Co^{2+} (pink)	A$^-$	O	A$^-$	S	*
Zn^{2+}	S	A$^-$,C	A$^-$,C	S	OH$^-$,NH$_3$
Mn^{2+} (pale pink)	S	A$^-$	A$^-$	S	—
Cu^{2+} (blue)	A$^-$,C	O	A$^-$,C	S	NH$_3$
Cd^{2+}	A$^-$,C	A	A$^-$,C	S	NH$_3$
Bi^{3+}	A	A$^+$,O	A$^-$	A$^-$	—
Hg^{2+}	D,O	O$^+$	A$^-$	S	—
Sn^{2+}, Sn^{4+}	A,C	A,C	A,C	A,C	OH$^-$
Sb^{3+}	A,C	A,C	A,C	A,C	OH$^-$
Ag$^+$	A,C	O	A$^-$,C	S	NH$_3$
Pb^{2+}	A,C	O	A$^-$,C	S	OH$^-$
Hg$_2^{2+}$	D,O	D,O$^+$	D,O	S	—

Key: S, soluble in water; no precipitate on mixing cation, 0.1 M, with anion, 1 M
  S$^-$, slightly soluble; tends to precipitate on mixing cation, 0.1 M, with anion, 1 M
  HW, soluble in hot water
  A$^-$, soluble in 1 M HC$_2$H$_3$O$_2$
  A, soluble in acid (6 M HCl or other nonprecipitating, nonoxidizing acid)
  A$^+$, soluble in 12 M HCl
  O, soluble in hot 6 M HNO$_3$
  O$^+$, soluble in aqua regia
  C, soluble in solution containing a good complexing ligand
  D, unstable, decomposes to a product with solubility as indicated
  I, insoluble in any common solvent
* Oxalates form many complex ions; oxides behave like hydroxides, but may be slow to dissolve; FeI$_3$ is unstable, decomposes to FeI$_2$ and I$_2$; CuI$_2$ is unstable, decomposes to CuI and I$_2$; AgBr and AgI do not dissolve in 6 M NH$_3$; HgI$_2$ is insoluble, but dissolves in excess I$^-$; Cr^{3+} and Co^{2+} can, under some conditions, form complexes with OH$^-$ and NH$_3$ respectively, but these complexes are not ordinarily produced under the conditions used in this text.

APPENDIX III

Some Anion-Anion Reactions

Most anions are stable in solutions containing other anions, but those that are oxidizing agents may react with those that behave as reducing agents. The rate of reaction between such ions usually depends markedly on both the acidity of the solution and the temperature. The conditions for reaction between some common ions in solution are summarized in the table below:

		Oxidizing Agents			
		NO_2^-	CrO_4^{2-}	ClO_3^-	NO_3^-
Reducing Agents	I^-	A (cold)	A (cold)	A (cold)	A (hot)
	Br^-	A (hot)	NR	A (hot)	NR
	S^{2-}	N (cold)	B (hot)	A (cold)	A (cold)
	SO_3^{2-}	N (hot)	A (cold)	A (cold)	A (cold)
	SCN^-	A (cold)	A (cold)	A (hot)	NR
	NO_2^-	—	A (cold)	A (cold)	NR

Key: A (cold), reaction occurs in cold 1 M HCl
 A (hot), reaction occurs in hot 1 M HCl
 N (cold), reaction occurs in cold solution at pH 7
 N (hot), reaction occurs in hot solution at pH 7
 B (cold), reaction occurs in cold 1 M NaOH
 B (hot), reaction occurs in hot 1 M NaOH
 NR, no reaction under any of the above conditions

Example. For NO_2^- and SO_3^{2-} the entry is N (hot). This means that if 0.1 M solutions of NO_2^- and SO_3^{2-} are mixed at pH 7 and heated in a boiling-water bath, an appreciable amount of SO_3^{2-} will be oxidized to SO_4^{2-} in about a minute. Since increased acidity greatly increases the rate of reactions of this sort, the reaction will also occur in cold, or hot, 1 M HCl, but not in cold or hot 1 M NaOH. Since lowering temperature decreases rate, the reaction will not occur to a significant extent in cold solution at pH 7, within a period of a few hours. At higher concentrations the reaction is more rapid. (In particular, 6 M HNO_3 is a much stronger oxidizing agent than 0.1 M NO_3^- ion in 1 M HCl.)

APPENDIX IV

Table of Atomic Weights (Based on Carbon-12)

	Symbol	Atomic No.	Atomic Weight		Symbol	Atomic No.	Atomic Weight
Actinium	Ac	89	[227]°	Mercury	Hg	80	200.59
Aluminum	Al	13	26.9815	Molybdenum	Mo	42	95.94
Americium	Am	95	[243]	Neodymium	Nd	60	144.24
Antimony	Sb	51	121.75	Neon	Ne	10	20.183
Argon	Ar	18	39.948	Neptunium	Np	93	[237]
Arsenic	As	33	74.9216	Nickel	Ni	28	58.71
Astatine	At	85	[210]	Niobium	Nb	41	92.906
Barium	Ba	56	137.34	Nitrogen	N	7	14.0067
Berkelium	Bk	97	[247]	Nobelium	No	102	[256]
Beryllium	Be	4	9.0122	Osmium	Os	76	190.2
Bismuth	Bi	83	208.980	Oxygen	O	8	15.9994
Boron	B	5	10.811	Palladium	Pd	46	106.4
Bromine	Br	35	79.904	Phosphorus	P	15	30.9738
Cadmium	Cd	48	112.40	Platinum	Pt	78	195.09
Calcium	Ca	20	40.08	Plutonium	Pu	94	[242]
Californium	Cf	98	[249]	Polonium	Po	84	[210]
Carbon	C	6	12.01115	Potassium	K	19	39.102
Cerium	Ce	58	140.12	Praseodymium	Pr	59	140.907
Cesium	Cs	55	132.905	Promethium	Pm	61	[145]
Chlorine	Cl	17	35.453	Protactinium	Pa	91	[231]
Chromium	Cr	24	51.996	Radium	Ra	88	[226]
Cobalt	Co	27	58.9332	Radon	Rn	86	[222]
Copper	Cu	29	63.546	Rhenium	Re	75	186.2
Curium	Cm	96	[247]	Rhodium	Rh	45	102.905
Dysprosium	Dy	66	162.50	Rubidium	Rb	37	85.47
Einsteinium	Es	99	[254]	Ruthenium	Ru	44	101.07
Erbium	Er	68	167.26	Samarium	Sm	62	150.35
Europium	Eu	63	151.96	Scandium	Sc	21	44.956
Fermium	Fm	100	[253]	Selenium	Se	34	78.96
Fluorine	F	9	18.9984	Silicon	Si	14	28.086
Francium	Fr	87	[223]	Silver	Ag	47	107.868
Gadolinium	Gd	64	157.25	Sodium	Na	11	22.9898
Gallium	Ga	31	69.72	Strontium	Sr	38	87.62
Germanium	Ge	32	72.59	Sulfur	S	16	32.064
Gold	Au	79	196.967	Tantalum	Ta	73	180.948
Hafnium	Hf	72	178.49	Technetium	Tc	43	[99]
Helium	He	2	4.0026	Tellurium	Te	52	127.60
Holmium	Ho	67	164.930	Terbium	Tb	65	158.924
Hydrogen	H	1	1.00797	Thallium	Tl	81	204.37
Indium	In	49	114.82	Thorium	Th	90	232.038
Iodine	I	53	126.9044	Thulium	Tm	69	168.934
Iridium	Ir	77	192.2	Tin	Sn	50	118.69
Iron	Fe	26	55.847	Titanium	Ti	22	47.90
Krypton	Kr	36	83.80	Tungsten	W	74	183.85
Lanthanum	La	57	138.91	Uranium	U	92	238.03
Lawrencium	Lw	103	[257]	Vanadium	V	23	50.942
Lead	Pb	82	207.19	Xenon	Xe	54	131.30
Lithium	Li	3	6.939	Ytterbium	Yb	70	173.04
Lutetium	Lu	71	174.97	Yttrium	Y	39	88.905
Magnesium	Mg	12	24.312	Zinc	Zn	30	65.37
Manganese	Mn	25	54.9380	Zirconium	Zr	40	91.22
Mendelevium	Md	101	[256]				

° A value given in brackets denotes the mass number of the longest-lived or best-known isotope.

APPENDIX V

Suggested Locker Equipment

2 beakers, 30 or 50 ml
2 beakers, 100 ml
2 beakers, 250 ml
2 beakers, 400 ml
1 beaker, 600 cc
2 Erlenmeyer flasks, 25 or 50 ml
2 Erlenmeyer flasks, 125 ml
2 Erlenmeyer flasks, 250 ml
1 grad. cylinder, 10 ml
1 grad. cylinder, 25 or 50 ml
1 funnel, long or short stem
1 thermometer (150°C)
2 watch glasses, 3 or 4 in
1 crucible and cover, size #0
1 evaporating dish, small
2 medicine droppers
2 regular test tubes, 18 × 150 mm
8 small test tubes, 13 × 100 mm
4 micro test tubes, 10 × 75 mm
1 test tube brush
1 file
1 spatula
1 test tube holder, wire
1 test tube rack
1 tongs
1 sponge
1 towel
1 plastic wash bottle
1 casserole, small